Journal of the International

anthrozoös

A multidisciplinary journal of the interactions of people and animals

Produced in cooperation with the Humane Society of the United States (HSUS), American Society for the Prevention of Cruelty to Animals (ASPCA), WALTHAM, and the International Association of Human–Animal Interaction Organizations (IAHAIO)

Editor-in-Chief

Anthony L. Podberscek
University of Cambridge
Department of Veterinary Medicine
Madingley Road
Cambridge CB3 0ES, UK
Ph: +44 - (0)1223-33 0846
Fax: +44 - (0)1223-33 0886
E-mail: alp18@cam.ac.uk

Associate Editors

Penny Bernstein
Department of Biological Sciences
Kent State University Stark Campus
6000 Frank Avenue
North Canton, OH 44720, USA
Ph: +1-330-499 9600
Fax: +1-330-494 6121
E-mail: pbernstein@stark.kent.edu

Patricia K. Anderson
Dept of Sociology & Anthropology
Western Illinois University
1 University Circle
Macomb, IL 61455-1390, USA
Ph: +1-309-298 1108
Fax +1-309-298-1857
E-mail: PK-Anderson@wiu.edu

Editorial Advisory Board

Karen Allen, State University of New York at Buffalo, USA
Frank R. Ascione, Utah State University, USA
Mara M. Baun, University of Texas Health Science Center at Houston, USA
Lisa Beck, Bryn Mawr College, USA
Andrea Beetz, Friedrich-Alexander-University Erlangen-Nuremberg, Germany
Brenda Bryant, University of California Davis, USA
Rebecca Cassidy, Goldsmiths College, University of London, UK
Stine B. Christiansen, University of Copenhagen, Denmark
Grahame Coleman, Monash University, Australia
Alan Costall, University of Portsmouth, UK
Beth Daly, University of Windsor, Canada
Clifton Flynn, University of South Carolina Upstate, USA
Erika Friedmann, University of Maryland School of Nursing, USA
Samuel Gosling, The University of Texas at Austin, USA
Kasey Grier, Winterthur Museum and Country Estate, USA
Sonya Hill, North of England Zoological Society, Chester Zoo, UK
Yuying Hsu, National Taiwan Normal University, Taiwan
Leslie Irvine, University of Colorado, USA
Garry Marvin, Roehampton University, UK
Ádám Miklósi, University of Eotvos, Hungary
Gene Myers, Western Washington University, USA
Jo-Ann Shelton, University of California, Santa Barbara, USA
Joanna Swabe, Independent Scholar, The Netherlands
Nicola Taylor, Flinders University, Australia

National Members of IAHAIO

Association Française d'Information et de Recherche sur l 'Animal de Compagnie [AFIRAC] (France)
Associazione Italiana Uso Cani da Assistenza [AIUCA] (Italy)
Belgian Association for Study & Information on the Human–Animal Relationship [ETHOLOGIA] (Belgium)
Delta Society (USA)
Delta Society Australia, Ltd. (Australia)
Drustvo Za Odgovoren Odnos Do Malih Zivali [DOMZ] (Slovenia)
Feleos Allatbaratok Egyesulete – Association of Responsible Pet Owners (Hungary)
Forschungskreis Heimtiere in der Gesellschaft – Pets in Society Research Group (Germany)
Human–Animal Bond Association of Korea [HAB Korea] (South Korea)
Institute for Interdisciplinary Research on the Human–Pet Relationship [IEMT] (Austria)
Institute for Interdisciplinary Research on the Human–Pet Relationship [IEMT] (Switzerland)
Japanese Animal Hospital Association (Japan)
Malaysian National Animal Welfare Foundation [MNAWF] (Malaysia)
MANIMALIS (Sweden)
Multidisciplinary Research Institute on the Relation Between Humans and Animals (The Netherlands)
Petcare Information and Advisory Service (Australia)
Polish Responsible Pet Owners Association [PRPOA] (Poland)
Research in Animal Therapy and Animal Assisted Education [RETAA] (Luxembourg)
Societá Italiana de Scienze Comportamentali Applicate [SISCA] (Italy)
Society for Companion Animal Studies [SCAS] (UK)
Society for the Study of Human–Animal Relations [HARS] (Japan)

Anthrozoös (ISSN 0892-7936 print; ISSN 1753-0377 online) is published four times per year by Berg Publishers, 1st Floor, Angel Court, 81 St Clements Street, Oxford OX4 1AW UK. Four issues form a volume.

2008 Subscription Rates

Print
Institutional (1 year): £145/US$270; (2 years) £232/$432

Online Only
Institutional (1 year): £123/$230; (2 years): £197/$367
Free online subscription for institutional print subscribers.
Full color images are available online.
Access your electronic subscription through www.ingentaconnect.com

2008 Membership Rates
Individual: £47.00/$89.00
Society Affiliates: £47.00/$89.00
Students/Unemployed Scholars: £23.50/$44.50
Corporate Membership: £141.00/$267.00
Lifetime Membership: £940.00/$1,780.00

Members and Affiliates of the International Society for Anthrozoology (ISAZ) receive the journal as part of their membership package.

Individual membership of the ISAZ is open to individuals currently or previously involved in conducting scholarly research within the broad field of human–animal interaction. Individuals who have an interest in the field of human–animal interactions, but who have not conducted scholarly research in the field, may apply to become Society Affiliates. For further details and to apply for membership, please see www.isaz.net.

Institutional Orders and Payment
Turpin Distribution handle the distribution of this journal. Institutional orders accompanied with payment (checks made payable to Turpin Distribution) should be sent directly to Turpin Distribution, Stratton Business Park, Pegasus Drive, Biggleswade, Bedfordshire SG18 8TQ, UK
Tel: +44 (0)1767 604951. Fax: +44 (0)1767 601640.
E-mail: custserv@turpin-distribution.com

Copyright © 2008 International Society for Anthrozoology (ISAZ), c/o Berg Publishers, 1st Floor, Angel Court, 81 St Clements Street, Oxford OX4 1AW, UK. All rights reserved. No part of the publication may be reproduced, stored in a retrieval system, or transmitted in any form or by any means, electronic, photocopying, recording or otherwise, without the prior permission of the publisher.

Indexing
Articles appearing in this journal are abstracted and indexed by Abstracts in Anthropology; Animal Behaviour Abstracts; CAB Abstracts; Current Advances in Ecological and Environmental Sciences; Current Contents/Social and Behavioural Sciences; Environmental Periodicals Bibliography; Focus on Veterinary Science and Medicine; Linguistics and Language Behaviour Abstracts; Psychological Abstracts; Referantivnyi Zhurnal: Biologiia; Science Citation Index; Social Science Citation Index; Sociological Abstracts; Veterinary Bulletin.

Berg Publishers is a member of CrossRef

Information for advertisers
Advertising orders and inquiries may be sent to:
Berg Publishers, 1st Floor, Angel Court,
81 St Clements Street, Oxford OX4 1AW UK
Tel: +44 (0)1865 241504, Fax: +44 (0)1865 791165
E-mail: enquiry@bergpublishers.com

Prepress production by Communicating Words & Images, Seattle, WA, USA.
E-mail: aptak5118@aol.com

Printed in the UK

Journal of the International Society for Anthrozoology

anthrozoös
A multidisciplinary journal of the interactions of people and animals
Volume 21, Issue 4, December 2008

CONTENTS

REVIEWS AND RESEARCH REPORTS — 317

Politics, Press and the Performing Animals Controversy in Early Twentieth-Century Britain
David A. H. Wilson — 317

Domestic Dogs as Facilitators in Social Interaction: An Evaluation of Helping and Courtship Behaviors
Nicolas Guéguen and Serge Ciccotti — 339

Attitudes and Actions of Pet Caregivers in New Providence, The Bahamas, in the Context of Those of Their American Counterparts
William J. Fielding — 351

The Relationship between Childhood Cruelty to Animals and Psychological Adjustment: A Malaysian Study
David Mellor, James Yeow, Norul Hidayah bt Mamat and Noor Fizlee bt Mohd Hapidzal — 363

Moral and Fearful Affiliations with the Animal World: Children's Conceptions of Bats
Peter H. Kahn, Jr., Carol D. Saunders, Rachel L. Severson, Olin E. Myers, Jr. and Brian T. Gill — 375

Comparison of Vegetarians and Non-Vegetarians on Pet Attitude and Empathy
Brooke Dixon Preylo and Hiroko Arikawa — 387

NEWS AND ANALYSIS — 397

New Books — 397
Conferences — 398

BOOK REVIEWS — 401

Bears: A Brief History — 401
Dog Behaviour, Evolution and Cognition — 402
When Species Meet — 405

INDEX TO VOLUME 21 — 409

WALTHAM®
THE WORLD'S LEADING AUTHORITY
ON PET CARE AND NUTRITION

Pioneers of research into the Human–Companion Animal Bond

ANNOUNCING the ISAZ-WALTHAM Research Award 2009

ISAZ and WALTHAM are delighted to announce a new grant program for the study of Human–Animal Interaction.

Details and forms are available on the www.ISAZ.net website.

Deadline for applications is January 1, 2009.

The WALTHAM Centre for Pet Nutrition, The science behind the Mars Petcare brands. For more information about WALTHAM science visit www:waltham.com.

Politics, Press and the Performing Animals Controversy in Early Twentieth-Century Britain

David A. H. Wilson
School of Humanities, University of Cumbria, Carlisle, UK

Address for correspondence:
David Wilson,
School of Humanities,
Faculty of the Arts, University
of Cumbria, Brampton Road,
Carlisle CA3 9AY, UK.
E-mail:
david.wilson@cumbria.ac.uk

ABSTRACT During 1921 and 1922, before the passage of the Performing Animals (Regulation) Act in 1925, a British parliamentary Select Committee undertook a detailed investigation into the degree to which animal performances in the circus and on the music-hall stage depended on cruelty. The investigation took place against a background of intense public interest that had been stimulated by the emergence of a new pressure group, the Performing Animals' Defence League. This paper examines the nature of political and press interests in the surrounding controversy and the detail of political involvement in a prolonged public dispute that suddenly arose after the war and was kept on the boil by fulminations in the national and trade press and on the floor of Parliament. As sides were taken, the motives, characteristics, and contributions of politicians and press are discussed, together with the debates contributing to the legislation and its aftermath.

Keywords: British history, performing animals, politics, press

 A controversy over performing animals had become consolidated in Britain soon after the First World War, largely through the Performing Animals' Defence League (PADL) and the associated work of the international Jack London Club, based on the appeal by Jack London in his foreword to *Michael, Brother of Jerry* (1917) to take direct action against animal performances. Following attempts by Joseph Kenworthy MP to bring this controversy to a head with the introduction of a prohibiting Bill in 1921, a Select Committee on Performing Animals was set up in the same year, and given clear terms of reference for its investigation.[1]

There had been some protest against performing animal acts well before the First World War, and such protests continued in the 1930s and later, after legislation in 1925 that many thought inadequate. Because sympathies among members of the Select Committee were almost equally

divided, and because the music-hall and circus professions mounted a campaign of defense of their interests equal to the agitation of the PADL and Jack London Club, the idea of legislative prohibition of performing animal acts was soon reduced to one of licensing, and the licensing provision of the resulting Performing Animals (Regulation) Act of 1925 has so far remained unaltered in the absence of any new legislation.[2] The incentive for further legislation was perhaps reduced because of the decline of the music hall with the rise of cinema and radio entertainment in the 1920s and 1930s, and because both the general public and local authorities began gradually to turn against animal performances: restriction of these has come to depend on policies adopted at local authority level. Animal acts also came to take up a smaller and less significant part in the British circus, and these acts started to become associated more with foreign circuses and trainers on tour in Britain. The Performing Animals (Regulation) Act of 1925 was therefore the first and last Act of its kind, in spite of subsequent attempts to amend or replace it.[3] The Act resulted from the weakening of a series of Bills and was not thought by critics of animal performances to reflect the findings of the Select Committee. For those defending the performing animals industry it was accepted as innocuous and therefore overtly supported. The development of performing animals legislation was reminiscent of that for vivisection in the 1870s. At that time, because of last-minute organized opposition from the medical establishment, the 1876 Bill was weakened and became in the same year an Act that not only protected vivisection through licensing but could also remove the requirement for anesthesia through exempting certificates (Ryder 1983, p. 135; Turner 1992, p. 209f).

The performing animals controversy was particularly appropriate material for the national press, with popular newspapers, especially, providing for readers the excitement of intense altercation between rival groups; pronouncements on a familiar form of entertainment about which ordinary people could be expected to take a moral view; the emotional stimulation of alleged cruelty to animals; and news of the possibility of government intervention. As far back as 24 August 1869, the *Times* had commented:

> The subduing of wild beasts … is merely the result of merciless thrashing when they are young. The application of the heavy cudgel, the iron bar or the red-hot ramrod on the tender limbs gives an impression which the threatening glance of him who wields these weapons keeps for ever fresh in the brute's memory. (cited by Turner 1992, p. 267)

In 1896, the *Daily Chronicle* attempted a systematic investigation of allegations of cruelty (Bensusan 1899, p. 118), and on 22 March 1899 the *Sketch* published claims by a first-hand witness of cruelty to performing bears. The *Observer* on 23 June 1912 included a letter from "Pom," who had been on the staff of one of the largest variety theatres for over seven years and referred to poor transport, lack of exercise, short-tempered and brutal foreigners, and cruel training devices. An alternative and more abstract view was expressed by a correspondent of the *Times* who believed that psychological suffering resulted from anticipation, memory, and the conscious association of moral significance to experiences; but an animal's awareness was "moment to moment" in nature, so the caged bird suffered less than the caged man. The correspondent argued that if it were thought acceptable to put animals to work, work for entertainment was also acceptable because all work for humans was contrary to animal instincts. In the meantime, cruelty in training for entertainment work would be counterproductive: the objectors were being anthropomorphic (Correspondent 1913. "Performing animals. The psychology of pain in man and beast," *Times*, 17 December).

Public concern about cruelty to performing animals (in capture, accommodation, transport, training and performance) was only articulated for the first time when the PADL was founded in 1914, and pressure increased on the founding of the international Jack London Club in 1918, following the author's publication of *Michael, Brother of Jerry,* focusing on the alleged cruelty behind animal performances.[4] In the previous century there had been much public interest, encouraged by the press, in the fate of two elephants, Chunee and Jumbo, but these were not performing animals, rather captive animal exhibits (Bondeson 2006, pp. 63–140). Their owners made them into individual animal "personalities," and their sad fates, each probably stemming largely from the effects of toothache, were accompanied by the promotion of highly profitable public anthropomorphism. Although their cases had nothing to do with the rights and wrongs of training animals for performance, the related debate about captivity and confinement was given a new impetus.

The performing animals controversy took its place in a succession of public animal welfare issues of the late nineteenth and early twentieth centuries that had so far centered on vivisection and then on the extravagant use of feathers in fashionable women's clothing. On 5 March 1917, Colonel Amelius Lockwood[5] had drawn attention in the House of Commons to the imminent arrival of ships containing hundreds of wild animals captured for show purposes, including very many apes and eight pandas ("very rare"), calling for the prohibition of such imports and asking which official was responsible for allowing the current importations. He was advised that no restrictions existed but that these would be considered (*Parliamentary Debates (Commons)* 1917–18, vol. 91, column 28). It is a reflection of the importance of this trade that it was maintained at a time when British and allied merchant shipping was suffering the worst of the U-boat campaign: approximately one million tons of such shipping were lost in April that year, and there was both public and official concern that the war could be brought to an early conclusion under unfavorable terms because of the lack of imported food and materials (Wilson 2001).

Kean (1998, pp. 179 and 191) has noted that the Humanitarian League, which concerned itself with human and animal interests alike, did not survive beyond 1919 because of the differences of opinion of its members on the recent war. Instead, some of its membership began to concentrate on the development of new campaigns, and a case in point was Ernest Bell and the PADL. Meanwhile, Methodists and Nonconformists continued to oppose animal cruelty inflicted for frivolous purposes, and took part in the criticism of animal performance in the 1920s. The First World War itself encouraged sober attention to the condition of humanity and its capacity for destructiveness, when service animals were recognized as victims, too, but a specific result of the war was a tendency to blame cruelty to performing animals on foreigners, especially Germans, as part-justification for a trade policy to boycott "alien acts" (Wilson In press a). After the war, there was more open discussion of emotive ethical issues such as this, and others, such as women's suffrage, sometimes shared support. As had been the case in earlier animal welfare issues, opponents of animal performance were characterized as middle-class or above, well-educated, especially in the arts, and often religious, and they were now supported by much of the national press as well as by means of campaigning groups, direct action and propaganda.[6] During the course of the controversy, the opposition of the groups on either side of it was sometimes also explained as reflecting a division of class as well as of perceptions of humane behavior or of legitimate economic interests, when it was proposed that privileged and hypocritical members of society forgot their interest in field sports in order to interfere with legitimate and harmless working-class entertainment. However, the daily press was generally critical of animal performance, and the alleged class factor seems not to have affected their editorial treatment: for example, the socialist *Daily Herald* was especially critical of animal acts. The *Era*, as a trade journal, complained:

> The daily press ranged themselves on the side of the objectors to animal performances, and on the eve of the debate in Parliament inflammatory articles with huge scare headlines were published in order to carry the day. (Editorial, "Looking back," *Era*, 1 March 1923, p. 13)

Those who resisted new legislation represented the commercial interests and claimed to defend innocent working-class entertainment from the "kill-joys" who they believed threatened a variety of freedoms without any evidence or justification (Wilson In press b). They acted through their trade organizations, such as the Variety Artistes' Federation (VAF), and expressed their views in their periodicals. Both sides were represented in Parliament and in the membership of the Select Committee.

The trade journals occasionally also presented the points of view of those opposing animal performances. "Nil Desperandum" wrote to the *Performer* (14 April 1921, p. 13) explaining his opposition: animals disliked performing and were not created to do abnormal tricks that gave them no benefit. The *Era* published a letter from E. J. Holland, editor of the *Anti-Vivisection Journal*, in which he criticized the increasing complexity of tricks, performing to timetable, and the stress of transport. He also supplied a campaign leaflet although the editor of the *Era* thought it contained no real evidence, but, rather, reference only to a band of gypsies and a dog left in a box at a railway station, and to common ailments (23 February 1921, p. 16). The *World's Fair* (11 June 1921), in one of its acknowledgements of the views of the opposition, reproduced a *Times* leading article on performing animals that asked why they amused, suggesting a psychological explanation was needed:

> There is something heartless, and even a little ill-bred, in our pleasure in performing animals ... forced to amuse us without being amused themselves. [The elephant was] a captive, a Samson making sport for the Philistines, who exult a little vulgarly in their triumph over her strength.

Use of animals in this way was a frivolous waste of man's valuable time, the article continued, and he did not know how to use his ingenuity properly. A genius like Darwin could spend months examining the natural behavior of earthworms, but the rest of us would only have interest in them if they could be taught tricks.

> If all animals did tricks by nature we would never notice them, but he [Darwin] would watch and watch to discover why they did their tricks; and he would get all his pleasure for nothing and win fame by it [without tyranny over the objects of his curiosity] ... The worms are not incommoded by Darwin; but we must humiliate even the king of beasts and our own national emblem before we can be interested in him.

Two years later (11 January 1923, p. 25), its columnist "Cosmopolite" wrote in the *Era* welcoming the idea of Royal Society for the Prevention of Cruelty to Animals (RSPCA) inspection and licensing, and to express sympathy with those who left their seats at unnatural animal performances of dogs, monkeys, or cats, because in Europe (as opposed to Britain) he had seen cruelty, remembering the "hidden hand" of the whip and the use of techniques such as water deprivation.

Kenworthy and Butcher

The two figures who played the most important part in promoting new legislation on performing animals were Joseph Kenworthy and Sir John Butcher, both of whom served on the Select Committee. Kenworthy (later 10th Baron Strabolgi) became MP for Hull Central in 1919, serving as a Liberal until 1926 and then as a Labour MP until 1931, having retired from the Royal Navy in 1920. He made the first move in Parliament on this issue, and had become a favorite victim of satire in the "Essence of Parliament" columns of *Punch*, where it was suggested that he asked too many questions as Supplementaries and broke procedural rules, and that he was a long-winded bore—over-confident, indiscreet and unrestrained—and had "bees in his bonnet."[7] His introduction of the Performing Animals (Prohibition) Bill provided a good opportunity for such sarcasm (see Figure 1):

> A Bill "to prohibit the exhibition of performing animals in places of public entertainment" was introduced by Lieutenant-Commander Kenworthy, and elicited thunderous cheers from all quarters of the House. The gallant gentleman, standing by the Speaker's Chair, was much affected by this testimony to his popularity. (*Punch*, vol. CLX, 16 March 1921, p. 213)

Like many of those who joined his cause and served on his side in the Select Committee, he saw no conflict between this interest and his enthusiasm for hunting and shooting. The suggestion of such conflicts was repeatedly made by those defending animal performance, and perhaps they helped to bring to the foreground an inconsistency which today would smack of hypocrisy. At the time, as Kenworthy and many others might have seen it, hunting was regarded as a natural, traditional, socially respectable, healthy, manly, unsentimental and character-building activity, sometimes involving the invigorating challenge of danger (McKenzie 2000). It should require knowledge and understanding of wildlife in its own context, and it was meant to be sporting, being based on codes, fair play, something approaching equal contest, and with little cruelty because of the quick despatch of the animal. It could also fulfil a diplomatic function abroad, when government representatives relaxed in the neutral but bonding environment of the field. In contrast, urban animal performances for working-class spectators were criticized for being unnatural, cruel, and undignified (for animal and human): a cowardly advantage was taken of a captive animal beneath a show of fake anthropomorphism and sentimentality, and children at the circus were liable to be deceived and corrupted by such spectacles. Kenworthy also objected to the humiliation of the circus lion because the animal was a national symbol.[8] Meanwhile, those who defended animal performance were clearly concerned for their livelihoods and objected in principle to the ignorance, prejudice, and spirit of interference they saw in their opponents, claiming their work to be legitimate and educational entertainment of importance to employment, the economy, and the very future of the music hall and circus.

In his autobiography, Kenworthy (1933) deals mainly with the navy and British foreign policy. There are also frequent and enthusiastic references to hunting expeditions abroad (see Figure 2), and his political concern for the welfare of performing animals (not itself mentioned at all in the autobiography) appears today as a contradiction that would not have been as well recognized at the time. Butcher, Bowyer, Colvin, and others who took Kenworthy's side over performing animals also shared his enthusiasm for hunting and shooting. However, those who wished to protect their profession from legislative interference lost no time in highlighting the contradiction, and perhaps in so doing inadvertently extended future attention to a wider range

A POPULAR TURN.
LIEUTENANT-COMMANDER KENWORTHY HAS PRESENTED A BILL FOR THE ABOLITION OF ANIMAL PERFORMANCES.

Figure 1. "A popular turn." Reproduced with permission of Punch Ltd, www.punch.co.uk

PANTHER SHOT IN NAWANAGAR STATE

Figure 2. Joseph Montague Kenworthy, hunting in India in 1926. From Kenworthy, J. M. 1933. *Sailors, Statesmen – and Others. An Autobiography*. London: Rich & Cowan. Reproduced with permission of 11th Baron Strabolgi.

of problems in human treatment of non-human animals. Kenworthy (1933, pp. 244 and 247) described hunting on elephant-back in India, in the foothills of the Himalayas in 1926:

> We shot anything which came along ... the bag comprised wild boar, wolf, four varieties of deer, a porcupine (which I wanted as a trophy), jungle cat, three varieties of partridges and quail ... I got a good tiger in Alwar State. He charged my elephant and then swerved off and I got in two .420 express bullets. [The Maharaja of] Alwar gave him two more .470s, and I changed rifles and hit him once more through the backbone with a .350 bullet. Each one of those five shots should have killed him. The first four were behind the shoulder, and my second bullet just missed the heart. Yet he ran back thirty yards, and then lay in a nullah. We walked him up with the elephants and found him stone dead, but he had charged back thirty yards ... His skin measures nine feet nine inches in length.

There is no doubt that Kenworthy expended much energy in drawing attention to the cruelties that he seemed to believe animals suffered in the course of training and performance, if not in hunting. However, given his character and tendency to self-advertisement, a cynic might wonder if he had decided to exploit the controversy for opportunistic political purposes. The press soon began to play its part in the debate, and Kenworthy's name appeared frequently. He told the *Sunday Chronicle* of the use of electricity to make performing animals dance and jump, and of heated irons to make wild animals snarl (cited in *World's Fair*, 5 March 1921, front page). Writing in the *Daily News* (cited in *World's Fair*, 27 August 1921, p. 5), he described the performances as degrading to the animal and the audience, and that long before the introduction of his Bill he had heard much from the entertainments profession about the subject. Complacency among his supporters, including the Prime Minister, meant that not enough had turned up to vote for the Bill, and its opponents then narrowly defeated it by introducing a licensing amendment that could have given immunity to licensed trainers from inspection by the RSPCA and the police. Kenworthy therefore withdrew the Bill, but because of the interest generated, the Select Committee was formed late in the parliamentary Session.

Shortly before the first meeting of the Select Committee, the *Daily Herald* ("Clear the stage," 3 June 1921) devoted an editorial to the subject:

> It is high time that the cruelty depended on the comedy of its clowns, not on the tragedy of its dogs and horses lashed, goaded, or starved into cutting dismal capers … No decent person can gain pleasure from seeing a wild beast tamed or a tame beast taught senseless tricks. Such things are an affront to Nature and an insult to the audience … We hold them in trust from Nature, and to exploit their intelligence and suffering for our profit is a contemptible breach of that trust. We are opposed to all exploitation, whether of human or non-human life. Let us clear the stage and clear our conscience too.

The following day's editorial decried the change in the Bill from prohibition to "ludicrous" licensing, and advocated the "easy and effective form of direct action" as practiced by the Jack London Club. The editorial was particularly critical of the "nauseous buffoonery" of Charles Stanton, Coalition National Democratic and Labour Party MP for Aberdare, during the Commons debate on the third reading, when he opposed the Bill. (Formerly a militant miners' leader, Stanton had abandoned the Independent Labour Party during the war in order to adopt a jingoistic pro-war stance and denounce many of his former colleagues for their pacifism.) On the same day the *Daily Express* published a critical article by Basil Tozer, a theatrical publicity agent who later served as a Select Committee witness (*Report from the Select Committee on Performing Animals, together with the Proceedings of the Committee and Minutes of Evidence.* London: HMSO, 1921, paragraph 2515f), entitled "How performing dogs are taught." The *World's Fair* later reported that the *Nation* (a weekly) regretted that this original Bill had been weakened ("Other People's Views," 17 November 1923, p. 12. The radical-Liberal *Nation* merged with the *New Statesman* in 1931).

Kenworthy's interest in this subject attracted as much hostility as support. The *World's Fair* published a letter from Captain Lloyd, an animal trainer for thirty years, who called him a crank, claiming there was great love between animals and trainers and referring to the case of the late "Lord" George Sanger. Sanger had told Queen Victoria of a lion that had died two days before his visit to Windsor, when he utterly broke down and the Queen, putting her hand on his shoulder, said, "The love between animal and trainer must be great indeed." Heated irons were used "to gull such asses as the Kenworthy click to make them think that the performance was a most dangerous one." Finally, to eliminate suspicion of cruel training methods, Lloyd challenged the press to witness the training of specially-bought new animals (*World's Fair,* 26 March 1921, p. 10). Other such challenges were also made directly to Kenworthy (e.g., from Leslie Walter and his horse and goat, *Era*, 30 March 1921, p. 12; and from George Dalmere, *World's Fair*, 4 June 1921, front page). An American trainer, Captain W. K. Bernard joined in the criticism, and suggested Kenworthy would be better engaged "if he looked into some of the affairs of poor children whose fathers gave up their lives possibly under his command" (*World's Fair*, 7 May 1921, p. 16). Nevertheless, in spite of this antagonism Kenworthy claimed later that in Parliament he served the interests of showmen, as when he obtained some fiscal concessions on taxes on their road transport; and he records great help in the election of 1929 from the Showmen's Guild and from the son of Pat Collins, the showman who had entered Parliament as Liberal MP for Walsall in 1922 and resisted new performing animals legislation (Kenworthy 1933, p. 279).

Figure 3. John George Butcher, Baron Danesfort. National Portrait Gallery, London. Ref. x66645. Reproduced with permission.

Figure 4. Sir Walter de Frece. Reproduced with permission of Laurence Asslinger-Hochschild.

Kenworthy's main ally in Parliament in opposing animal performance was Sir John Butcher, Conservative MP for York, created Baron Danesfort in 1924 (see Figure 3). He was less colorful than Kenworthy but shared his attitude to animals. The *Times* recorded:

> He was fond of hunting and shooting, and, like many another sportsman, where animals were concerned he was always their warmest champion; he was vice-chairman of the RSPCA, and an ardent opponent of vivisection … All his life he had been a sportsman. He shot and rode well into old age.

He was described as a diehard Tory who earnestly maintained the most extreme Ulsterman's attitude and was no opportunist, being unswerving, of passionate sincerity, outspoken, and having honest conviction (Obituary, Lord Danesfort, *Times*, 1 July 1935).

Of those in Parliament who defended the professional interests of the circus and music hall and who attempted to limit any new legislation, leading parts were played by Sir Walter de Frece (see Figure 4), Captain James O'Grady (see Figure 5), and Charles Jesson. De Frece was married to Vesta Tilley, the actress and male impersonator. He was Unionist MP for Ashton-under-Lyne between 1920 and 1924, for Blackpool between 1924 and 1931, and managing director of a company controlling up to twenty theatres. O'Grady, Labour MP for Leeds South-East (1918–1924) and formerly secretary of the National Federation of General Workers, had addressed a letter to every member of the House of Commons on the performing animals matter. He had helped found the VAF, and in "Triumph of the Animal Trainers" he was described as a great friend of the music-hall profession (*Encore*, 9 June 1921, p. 3). Jesson was Coalition National Democratic and Labour Party member for Walthamstow West between 1918 and 1922, and late organizer of the Musicians' Union.

De Frece wrote to the *Daily Telegraph* against the Bill (*Performer*, 14 April 1921, p. 9), and on the commencement of Standing Committee "D," he gave all the current reasons for

Figure 5. Sir James O'Grady when Governor of Tasmania, 1924 to 1930. La Trobe Picture Collection, State Library of Victoria. Reproduced with permission.

Figure 6. Trevelyan Thomson. National Portrait Gallery, London. Ref. x122078. Reproduced with permission.

opposing it (*World's Fair*, 16 April 1921, front page). A letter in the *Daily Herald* from Ernest Bell, chairman of the PADL, referred to Jesson's challenge to witnesses in sympathy with the PADL to substantiate their claims about cruelty under cross-examination before music-hall managers and trainers at the House of Commons on 31 May 1921 in Committee Room 6. This challenge was accepted, but the witnesses were shown into Committee Room 12, which was empty, and they were refused entry into Room 6 itself, as if there had been a change of heart over the feasibility of disproving cruelty (*Daily Herald*, 2 June 1921, p. 4). The explanation given was that an MP objected to Bell's entrance, perhaps because of lack of time (*Era*, 15 June 1921, p. 14). The *Performer* (2 June 1921, p. 8) reported this meeting of MPs, organized by O'Grady at the request of the VAF, about the possible government Committee of Inquiry into the training of animals, the result, it said, of "hypersensitive and morbidly sentimental humanitarians." It attributed the subsequent failure of the Bill, against which O'Grady had acted on behalf of the VAF, to flimsy evidence for cruelty (*Performer*, 9 June 1921, p. 5). But it now warned of the forthcoming Select Committee and the legislation that might result. The VAF arranged once more for O'Grady to watch its interests (*Performer*, 16 June 1921, pp. 5 and 9).

Investigation, Debate and Legislation

The Performing Animals (Regulation) Act came into effect on 1 January 1926, following protracted investigation and debate, both in Parliament and in the press. Earlier related legislation consisted of the Wild Animals in Captivity Protection Act, 1900, and the Protection of Animals Act, 1911, drafted by the RSPCA and introduced by George Greenwood MP, one of its council members. Under the 1900 Act it became illegal to cause any unnecessary suffering, or to cruelly abuse, infuriate, tease or terrify; and the 1911 Act made illegal, again by cause or permission, similar mistreatments of any animal, including those of a more everyday nature, such as beating, overloading, or stressful transportation, and also staged animal fighting and baiting, poisoning and cruel operations. Kenworthy had originally raised the subject of performing animals in Parliament on 21 February 1921, asking the Prime Minister if he was aware of the …

> … many proved cases of cruelty involved in the breaking and training of wild and domestic animals, other than horses, for performances on the stage and in travelling circuses, and that an increasing number of people object to these performances.

He asked if the Prime Minister would consider introducing legislation on this subject. The response was that the 1911 Act offered sufficient protection (*Parliamentary Debates (Commons)* 1921, vol. 138, column 545). Kenworthy thereupon promptly presented the Performing Animals (Prohibition) Bill, supported by, among others, Trevelyan Thomson (Independent Liberal member for Middlesbrough West from 1918, see Figure 6), who would also later serve on the Select Committee. The Bill sought to prohibit the exhibition of performing animals and birds in places of public entertainment. A Standing Committee amendment allowed for the exhibition of horses and of animals ordinarily used in military tournaments, and the exhibition of animals "by any person licensed to train and exhibit animals by any Justice of the Peace in such form as may be prescribed by a Secretary of State" (*Report from Standing Committee D on the Performing Animals (Prohibition) Bill with the Proceedings of the Committee.* London: HMSO, 1921). Although Kenworthy withdrew his weakened Bill, as a result of debate in Parliament it was decided to appoint a Select Committee and nominate members (*Parliamentary Debates (Commons)* 1921, vol. 144, column 1243). At this time, *Punch* (vol. CLX, 23 March 1921, p. 229) contributed to the controversy with a sympathetic poem by "Algol" about a hippopotamus captured and transported to the indignity of the stage, "To A Performing Hippopotamus":

> … What dost thou here, majestic river-horse,
> Where airs are cold and audiences coarse?
>
> … He must have been a most repulsive brute
> Who marked thee down as profitable loot
>
> … And here thou art! The punctual curtain falls;
> Coldly thou tak'st a brace of well-earned calls,
> Dost off the motley and reseek'st repose,
> Sunk in thy tank – and boredom – to the nose.

The Select Committee was appointed on 12 July 1921, and at the conclusion of its work that Session, a Report with minutes of evidence had been brought up under "oral answers," read, and ordered to be printed; but it was pointed out that the inquiry was not completed (*Parliamentary Debates (Commons)* 1921, vol. 146, column 657). Soon after, Brigadier-General Colvin asked the Prime Minister whether the Select Committee, which had taken a large amount of important evidence at the end of the last Session, would be reappointed forthwith so that the evidence might be completed and the Committee enabled to issue its Report (*Parliamentary Debates (Commons)* 1922, vol. 150, column 605). The Secretary of State for the Home Department replied in the affirmative, and the minutes of evidence of the Select Committee of the first Session of 1921 were later referred to the reappointed Select Committee (*Parliamentary Debates (Commons)* 1922, vol. 150, column 2065).

As was usual, the Select Committee was empowered to send for "persons, papers and records," through an informal invitation by means of press notices to individuals to give personal evidence or to interested organizations and individuals to send memoranda on the subject of the enquiry. The Select Committee, being an extension of the House, possessed substantial powers, greater than in a court of law, to require answers to questions and the production of documents. The members of the Select Committee approached witnesses according to their own prejudices or preconceptions. A witness was required to answer questions and to be truthful. Evidence was not given on oath, therefore punishment for contempt rather than perjury would apply to giving false evidence. Witnesses were protected from the consequences of giving evidence, such as slander. According to convention, members of the Select Committee co-operated and did not unreasonably interrupt during the investigation, although a question put to a witness could be objected to by a member, much like the interaction of prosecution and defense in a court (Limon and McKay 1997, pp. 646–54).

With a quorum of five, the Select Committee comprised: Captain George Bowyer, Conservative member for Buckingham between 1918 and 1937, when he was created 1st Baron Denham of Weston Underwood; Sir John Butcher; Brigadier-General Sir Richard Beale Colvin, Unionist member for Epping between 1917 and 1923 (chairman); Sir Walter de Frece; Lieutenant-Colonel Sir Raymond Greene, Conservative member for North Hackney between 1910 and 1923; Mr Charles Jesson; Lieutenant-Commander Kenworthy; Captain James O'Grady; Mr Alfred Raper, Coalition Unionist member for East Islington between 1918 and 1922; Mr Frederick Roberts, Labour member for West Bromwich between 1918 and 1931, and 1935 and 1941; Mr James Seddon, Coalition Labour member for Hanley between 1918 and 1922 and chairman of the Trades Union Council in 1914; Mr Charles Stanton (the only nominee who subsequently failed to take part); Mr John Swan, Labour member for Barnard Castle between 1918 and 1922; Mr Trevelyan Thomson; and Lieutenant-Colonel Claud Willoughby, Unionist member for Rutland and Stamford between 1918 and 1922. Twenty witnesses were examined over six days. As recommended, the Select Committee was re-appointed on 22 February 1922 to continue its work. It heard evidence from a further twenty-three witnesses over six days and met twelve times between 2 March and 9 May, 1922 when a draft Report was produced.

Amendments to the draft were extensively argued over and so further meetings were held on 11, 16 and 18 May before confirmation of the Report (*Report from the Select Committee on Performing Animals, together with the Proceedings of the Committee and Minutes of Evidence.* London: HMSO, 1922). In each division over the amendments, the committee members were nearly always almost evenly split according to their differing views on the desirability of new legislation. Broadly, the views reflected on the one hand those, some representing the humane societies, who believed that cruelty was an existing problem, and on the other those, some representing professional interests, who believed it was not. Those eager for legislation were in the slight majority, and amendments were therefore made that resulted in a final Report that was a little more powerful than the draft. These amendments served to emphasize that ill-treatment was still common; removed reference to benefits to children from any animal performances; suggested that cruelty had to be obvious to spark public disapproval; and enabled inspection without previous notice given, and the prohibition both of performances of animals thought to be cruelly trained abroad and of conjurors' devices involving cruelty. It was now also recommended that training and performance of anthropoid apes be prohibited. An amendment from Sir John Butcher to include larger carnivora had been narrowly defeated, but it was agreed that they should have the special attention of the proposed Committee of Supervision. Dissatisfaction with the sum of the amendments in the final Report on the part of those opposed to legislation led them to vote against its submission to the House of Commons in amended form, but they were defeated by five votes to seven.

The Report includes an account of the nature of evidence given and the Committee's reaction to it, a summary including a general expression of opinion, and recommendations. The Committee reported that cruelty existed, that the public was opposed to it, and that the profession was willing to co-operate against it. Evidence covered the previous 25 years, but difficulty was experienced in obtaining evidence of cruelty from persons who considered that their chances of employment might be endangered. The Committee received a large number of letters as well as sworn declarations, but this evidence was of less value because it could not be made subject to cross-examination. The Committee remarked that its enquiry had drawn press attention, and audiences of animal acts had become very critical as a result. A critical public was thought to be the animals' only existing source of protection, since there was no legal right of inspection of practices. It was noted: "It is not contended that admission [for inspection] would be denied ..., but a warning might be given that would render such a visit useless." There was also a need for closer inspection of transport and accommodation arrangements, and it was felt, too, that animals trained for use in film productions should benefit under the proposed measures. Two categories of animals were identified, "savage and untrustworthy" and "more domesticated and docile:" cruelty to the first could be involved especially in subjugation and to the second in training for tricks. The effect on animals of performance to a timetable was thought to vary according to their temperament. Finally, the Committee stated that in 1914, about 20% of trainers had been British, and they were considered more humane.

The Committee considered that wholesale prohibition was not necessary, but that supervision was needed for trainers, training establishments, and animal performances, and that existing penalties were inadequate. It recommended that the proposed Committee of Supervision should have powers to prohibit, restrict, suspend, and modify performance or training, including prohibition of acts resulting from suspect training methods abroad. Trainers and training establishments should be registered, and local authority, RSPCA, and police officers should have right of access at all times, without previous notice given, to performances and training establishments. The use of cruel appliances by conjurors, and the training and performance of anthropoid apes should be prohibited, and special attention should be given by the Committee of Supervision to the larger carnivora. Finally, stiffer penalties should be applied to those convicted of cruelty. The Report of the Select Committee therefore marked the point at which the possibility of early general prohibition was replaced by an intention to regulate. A leading article in the *Times* of 4 December 1922 commented:

> The Report contains more than sufficient grounds for immediate legislation ... we accept, although with reluctance, the decision that the time has not yet come for total prohibition of this particular form of exploitation of animals for the pleasure of men.

It also disagreed with the distinction made by the Select Committee between wild and domestic animals (giving special attention to the circumstances of the larger carnivora), because domesticated or docile ones were at least as susceptible to cruelty.

Following the work of the Select Committee, whose final Report was published in May, 1922, a new Bill to regulate the exhibition and training of performing animals was presented in 1923 by Colvin and supported by Butcher, Thompson, Greene, Bowyer, Roberts, and Kenworthy. It called for the registration of trainers and exhibitors of performing animals; power for the Secretary of State by regulations to prohibit or make conditions on the use of apes, (and

Figure 7. Sir Richard Beale Colvin. National Portrait Gallery, London. Ref. x166703. Reproduced with permission.

Figure 8. Hugh Cecil Lowther, 5th Earl of Lonsdale. National Portrait Gallery, London. Ref. x85187. Reproduced with permission.

now also) large carnivora, and appliances for conjuring tricks; power of a court order to prohibit or restrict an activity; power of a police constable or representative of an approved humane society to inspect premises "at all reasonable times" (the Select Committee had recommended such inspections "without previous notice given"); possible deregistration following conviction; and an Advisory Committee for the Secretary of State concerning regulations, consisting of a chairman with two humane society members and two from the profession (instead of the staffed Committee of Supervision proposed by the Select Committee, which would have been costly, requiring offices).

During the debate of 3½ hours on the second reading on 23 March 1923 (*Parliamentary Debates (Commons)* 1923, vol. 161, columns 2961–3020), Colvin (see Figure 7) said that legislation was necessary because evidence had been presented that there was cruelty, that job losses would only affect the cruel, and that the public needed reassurance about such entertainment, which was becoming less attractive. Referring to the findings of the Select Committee, he noted that some animals such as seals were suitable for training without cruelty or intimidation, while others, such as large carnivora and apes, were not. There was also the problem of inept use of conjurors' apparatus, such as in the disappearing-bird trick. At the same time, he was against the "frivolous" interruption of performances by protestors, and would support a special amendment to suppress them. O'Grady blocked the Bill on behalf of the VAF (*Performer*, 28 February 1923, p. 5) and moved for its rejection, claiming that the Select Committee had been given no proof of cruelty in evidence, resulting in an almost even split of members when asked to sign approval of the Report. He said the Bill would effectively result in an Act of prohibition unless exemptions were made by the Home Secretary, who would feel uncomfortable because of the behavior of vindictive societies, as in the case of Lord Lonsdale (see Figure 8).[9] There had been few recent convictions, and he claimed that cruelty before the War had usually involved Germans (often blamed and referred to during this controversy as the "alien enemy;" Wilson In press a). Representing the VAF and being one of its originators, he confirmed that the organization would expel a member for cruelty, that the Bill would destroy livelihoods, and that the RSPCA already had full access for inspection. He attempted to pull apart the evidence of some Select Committee witnesses, saying that the Bill was promoted by irrational prejudice.

Major Thomas Paget, Conservative MP for Bosworth, questioned the scope of performing animals as understood in the Bill, and the interference of inspections. As a warning, Major Leonard Molloy, Conservative MP for Blackpool, cited the RSPCA's president at the annual meeting in 1921 who considered the Bill of that time as a step towards the total abolition desired by some, and Molloy also criticized undesirable direct-action tactics.[10] He regarded the current Bill as an assault on the working class and its entertainment, and felt that general sympathy with the movement towards humane treatment of animals could be lessened as a result. An extension of the law was not necessary, and the Select Committee recognized the opposition of the profession to cruelty. The Bill was "the attempt of a noisy, but socially influential, class to interfere with the legitimate amusements of the people," resulting from the "fulminations of a lot of hysterical men and women" and the "senile arguments of a lot of old women, of both sexes." But George Buchanan, Labour MP for Glasgow Gorbals, believed the Bill should go further and include other amusements such as steeple-chasing, fox-hunting and human performances. Employment and the cinema industry should not have to depend on cruelty, he said, and supervision was necessary because professionals placed popularity and profit above animal welfare.

Sir Walter de Frece, speaking for the Society of West End Managers, the Theatrical Managers' Association, the VAF, circus interests, and the entertainment world, asserted that it was only a Private Member's Bill because the government did not believe in the report of the divided Select Committee [which nevertheless acknowledged as a whole that cruelty existed], and it reminded him of the Puritans who were not concerned by animal cruelty but by human pleasure. Unemployment would result because exhibitors would not want the bother of inspection. Although he did not object to registration and honest investigation and inspection, he objected to inquisition by people who would "have to make a case to keep their jobs, and who are not honest critics of what they see," leading not just to sentimental interference but also to widespread state control through the autocratic regulatory powers given to the Home Secretary via the Bill's proposed Advisory Committee. Butcher said that the Select Committee did hear evidence of cruelty, and all had agreed to the reference to it in the Report, including Sir Walter de Frece. He commented that "under the existing law, the facilities for inspection, obtaining evidence, prosecution, and so on, were insufficient to stop this cruelty," and that training took place in secret, making cruelty easier and inspection necessary. What animals to include (e.g., as mentioned in the debate, Grand National horses, or snakes), could be decided on at Committee Stage, and the idea of wholesale prohibition was a bogey of no consequence. He also noted that 80% of trainers were foreign, many being Germans whose cruelty produced what they called good results, and he claimed:

> You do not get rid of the strong determination in this country to suppress every form of cruelty towards animals—a determination which exists both in this House and very largely throughout the country, and not least amongst the newly enfranchised voters, namely, the women—…by calling it sentimentalism.

Bowyer admitted that it was very difficult to expose evidence of cruelty, which was often caused by animals being forced to perform on demand and to a timetable, as explained by

Figure 9. Sir George Kynaston Cockerill. National Portrait Gallery, London. Ref. x121362. Reproduced with permission.

Figure 10. "Will it hatch?" Reproduced with permission of Punch Ltd, www.punch.co.uk

Peter Chalmers Mitchell, whom he thought the most important witness examined by the Select Committee.[11] Benjamin Tillett, Labour MP for Salford North and an ex-circus boy, thought training was only possible by kindness, and cruel people were ostracized by the circus, where chivalry was shown to animals. He asked why that day's Grand National, or hunting, or slums, were not the object of concern: this was an attempt at class legislation. Sir Walter Greaves-Lord, Conservative MP for Norwood, asked what exactly would be the grounds for objection leading to a conviction, and was worried by the subjectivity of magistrates, an appeal against whose decision would mean interim loss of earnings, and expense. Thomas Groves, Labour MP for Stratford West Ham, believed that cruelty was involved in unnatural performances, and that legislation in a civilized society necessarily interfered with individual liberty. William Bridgeman, Conservative MP for Oswestry and Secretary of State for the Home Department, said that the government would not use its parliamentary whips in the vote, but that the Home Office welcomed the Bill, and opposition to it that day had been weak. There was strong public feeling against cruelty to animals, and those in societies opposed to cruelty to animals were also in societies opposed to cruelty to children. Stephen Walsh, Labour MP for Ince, saw no evidence that the 1911 Act had not had effect, but approved of the large attendance in the House as a reflection of British society's wish to reduce unnecessary animal suffering. James Gould, Conservative MP for Cardiff Central, said that everyone had been badgered by societies, but that only an increase in penalties was necessary. The Bill then passed its second reading and was put to Standing Committee. Subsequent amendments by Standing Committee removed the proposal for the Secretary of State's regulatory powers and the associated Advisory Committee; and allowed for inspection only by a police constable or a local authority official, neither of whom could go on or behind the stage during a performance.

In the next Session of Parliament, Kenworthy asked the Under-Secretary of State for the Home Department whether the government would introduce this Bill, as amended (as Bill no. 208) to meet the views of all as a compromise, which would have passed through the last Parliament as an agreed measure had not time run out because of Dissolution. Rhys Davies replied that the Government was now too busy and had no time, but that if agreement had been reached in the last Session he hoped a Private Member's Bill might be successful (*Parliamentary Debates (Commons)* 1923, vol. 169, column 1187). The Bill (no. 81) was duly presented in 1924 by Brigadier-General Sir George Cockerill, Conservative MP for Reigate (see Figure 9), and supported by former Select Committee members Bowyer and Kenworthy, and by Sir Sydney Henn (see Figure 10). It failed to compete successfully with other business before Dissolution, but it had been debated and had passed its second reading in the Lords on 3 April 1924. In that debate Lord Raglan disapproved of police powers of inspection without warrant, while Viscount Knutsford, describing himself as experienced in training animals and also a conjuror by profession, expressed dislike of the way of life of performing animals rather than the nature of their training; and he felt there was less of a problem with cruelty than with dignity (*Parliamentary Debates (Lords)* 1924, vol. 57, column 147ff):

> Personally, I dislike very much seeing beautiful animals, either day after day or night after night, in uncongenial and unnatural surroundings, performing tricks, and I am always struck when I see them with the much greater dignity that there is in the animals than in the man whose commands they obey.

In the Session of 1924–25, the Performing Animals (No. 2) Bill was presented as a Private Member's Bill by Cockerill and supported by Select Committee members Bowyer, de Frece, Kenworthy, and Thomson. It was substantially the same as the latest form of the previous Bill, except that registration was now made specific to the exhibition of performing animals at any entertainment to which the public were admitted, whether on payment of money or otherwise, and to training for such exhibition; and the Act would not apply to the "training of animals for *bona fide* military, police, agricultural, or sporting purposes, or for the exhibition of any animals so trained." The Bill then still remained substantially the same after amendment by Standing Committee and Lords Amendments (*Parliamentary Debates (Lords)* 1925, vol. 61, columns 254ff, 473ff and 528ff), although only public, rather than RSPCA or private, potentially "vexatious," prosecutions would be allowed. Invertebrates were excluded, thereby denying benefit to the performing flea—*de minimus non curat lex*—and the possible reasons for the decision on fleas were offered by Cockerill in Consideration of Lords Amendments (*Parliamentary Debates (Commons)* 1923, vol. 185, column 967f). The Bill was now passed as the Performing Animals (Regulation) Act.[12] During the Lords Debate on its second reading, Lord Raglan (see Figure 11) had again taken the opportunity of expressing his disapproval with it and describing its origins (*Parliamentary Debates (Lords)* 1925, vol. 60, column 932ff):

> Some four or five years ago a number of hysterical women suffragettes, being in search of an emotional appeal and having no excuse for burning down buildings or molesting the noble Earl [perhaps Lonsdale] opposite, formed themselves into bands and went round to circuses, where they howled and generally made themselves objectionable. Eventually they succeeded in getting the sympathy of the noble Lord [probably Danesfort], and other members of the House of Commons and a Committee was set up … It would seem that this Bill is required, not so much to protect the animal from the cruelty of the trainer as to protect the trainers from the cruelty of Lord Danesfort and his friends …

Figure 11. FitzRoy Richard Somerset, 4th Baron Raglan. National Portrait Gallery, London. Ref. x122446. Reproduced with permission.

Unfinished Business

In the 1930s, during the aftermath of the controversy and of the resulting legislation, the level of demand for animal acts was discussed in the House of Lords. Lord Jessel reported that although Bertram Mills had dropped animal performances in 1930, public demand induced him to reintroduce them the following year (*Parliamentary Debates (Lords)* 1933, vol. 87, column 730). On the other hand, Lord Chesham referred to a Home Office report stating that in 1929 there were 206 British trainers, but by 1933 the figure for trainers in England was 113, and Lord Danesfort claimed that the number of trainers of scheduled animals had gone down considerably, reflecting a decline in demand (*Parliamentary Debates (Lords)* 1933, vol. 87, column 737). The 1925 Act has not been superseded, but unsuccessful Bills of amendment were soon forthcoming after its passage. The first was the Performing Animals (Regulation) Amendment Bill of 1930. It was presented in the House of Lords by Danesfort and then brought to the House of Commons, and proposed no exhibition or training for performance of any ape, big cat, or hyena, exemption applying to the Zoological Society of London or any other organization which had as its principal object the exhibition of animals for educational or scientific purposes.

A similar Lords Bill, now including bears, and devised largely by the RSPCA of which Danesfort was vice-chairman (as well as being a leading figure in the National Canine Defence League), was presented by him in 1933, to prevent cruelty to the specified animals (usually trained abroad) and danger from them to their trainers and exhibitors. He cited an extract from the *Times* of 3 April 1933 by its Berlin correspondent, noting that the authorities there (perhaps, in retrospect, for more than one reason) had banned bear performances led by gypsies or other itinerants, because of the plight of the animals. Danesfort hoped the Bill would strengthen the 1925 Act, which had been emasculated through compromise, and he called in support the original evidence given by Chalmers Mitchell to the Select Committee, and to recent cases of human injury. In 1931 the Home Secretary and the President of the Board of Education had been asked by Peter Freeman, Labour MP for Brecon and Radnor (and later president of the Vegetarian Society between 1937 and 1942), to consider legislation to prevent children attending animal performances, and reference was made to the recent unanimous resolution of the Royal Society of Teachers (*Parliamentary Debates (Commons)* 1930–31, vol. 244, column 1676; vol. 247, column 615f). In 1933 Lord Banbury of Southam, during the second reading of Danesfort's Bill (*Parliamentary Debates (Lords)* 1932–33, vol. 87, column 733), said that it was

> ... far better that the children should be brought up to consider kindness to animals than get their amusement from seeing those animals forced to do things contrary to their nature ... [the Bill] may do some good to children.

Danesfort referred to a mistaken statement from the Home Office that there had been no cases made out for cruelty to performing animals since the Act of 1925, when there had been recent convictions (not for training but for exhibition and management) for the death of a foreigner's lion as a result of a "wall of death" display, and for a case of close confinement of lions. But there was a lack of convictions under the Act because training was done abroad or behind the scenes, out of the reach of the RSPCA inspectors, who had no legal power of inspection. Furthermore, Danesfort did not think the police and local authorities exercised their powers much to carry out those inspections permitted when a performance was not taking place. Meanwhile, there had been a slight increase in the training and exhibition of lions, tigers, and bears since 1930, and there had been many recent cases of fatal injuries sustained by trainers who were usually foreign. On behalf of the Home Office, Lord Lucan replied that since the 1925 Act no further evidence of cruelty had been brought to the notice of the Department, including the two lion cases, perhaps explained by a meeting of chief constables in 1928 and a related systematic inspection of exhibitions in the boroughs when no concerns arose. Police inspection methods and results were checked in 1929 (*Parliamentary Debates (Lords)* 1932–33, vol. 87, columns 723f and 741f). During this second reading of the 1933 Bill, Lord Carnock attempted to downplay arguments about unnatural behavior by claiming that he had seen chimpanzees enjoying a whisky and soda. He qualified concerns for animal dignity (*Parliamentary Debates (Lords)* 1932–33, vol. 87, column 740f):

> It may be undignified for a lion to ride round a circus ring on the back of a horse, but it is much more undignified for a poor man with his wife and children to go from doss house to doss house unemployed and penniless, and that is what this Bill would mean.

Lord Jessel felt that the Bill should not be passed without the investigation of another Select Committee (*Parliamentary Debates (Lords)* 1933, vol. 87, column 719ff). Toole Stott (1971, p. 63) claims that Bertram Mills helped to defeat this Bill by asking him to hire a Fleet Street photographer to accompany him to London Zoo in the guise of Australian tourists, so that photographs could be taken of animals performing tricks: these were then distributed to Members.

Another similar Bill was presented in the Commons by Daniel Frankel, Labour MP for Mile End, in 1937, to prohibit the use of dogs, seals, monkeys, apes and baboons, or any other animals imported into Great Britain of a greater age than six months. The Bill provided for rights of inspection of training premises under the 1925 Act by representatives of recognized humane societies (the RSPCA, Our Dumb Friends' League, or others approved as having similar objectives by the Secretary of State) who could also be complainants; and they should be allowed to attend performances and exhibitions of performing animals for the purpose of inspection (if necessary and without interference on or near the stage), with or without previous notice, when apparatus and materials could also be inspected. Frankel's measure would also make the particulars of registration and court orders under the 1925 Act and this proposed Act more readily available for public inspection. The aims of the Bill were to ensure that performing animals were born in Britain and to improve the machinery of legislation. Penalties could for the first time include imprisonment, and confiscation and care of animals by a humane society, or, if necessary, destruction; and the possibility of permanent disqualification. Finally, each animal would be identified individually at registration. The same exemptions applied as those included in the 1929–30 Bill.

Before the outbreak of war, an unsuccessful Bill much like Danesfort's of 1933 was presented in the Lords by Charnwood in 1938. He had written to the *Times* (19 August 1937, p. 8) to protest at the confinement of large, active beasts, even if they appeared healthy. In a "good" menagerie, two to four lions had a space of 24 × 9 or 10 feet, sometimes divided into two by a partition, never having a wider range of movement. Feeling repugnant about this did not result from "feeble sentimentality," and the question needed public attention. John C. Lockwood, Conservative MP for Hackney Central between 1931 and 1935, replied that these feelings were shared widely. In the last Parliament, the Home Secretary had told him that existing legislation was sufficient, but mental cruelty needed to be considered, too (*Times*, 23 August 1937, p. 15). A little later, Sir Robert Gower (Conservative MP for Gillingham between 1929 and 1945, chairman of the Animal Welfare Committee in the House of Commons, and chairman of the RSPCA between 1928 and 1951), one described as opposing all wild animal captivity except in certain, special near-ideal conditions, wrote that the dangers of the natural state were better than imprisonment (*Times*, 26 August 1937, p. 8):

> Happiness is perfect harmony between function and environment, and a truly civilized community will not take enjoyment in seeing wild animals leading empty, unnatural lives.

Acknowledgements
I am grateful for help during the preparation of this paper from Laurence Asslinger-Hochschild; Mike Casselden; David Kenworthy, Lord Strabolgi; Patrick Newley; and Christopher Woodward.

Notes
1. The Select Committee's brief was "to inquire into the conditions under which performing animals are trained and exhibited, and to consider whether legislation is desirable to prohibit or regulate such training and exhibition, and, if so, what lines such legislation should follow" (*Parliamentary Debates (Commons)* 1921, vol. 144, column 1243). Reports were made after two hearings. It met seven times between 19 July and 11 August, but did not manage to conclude its investigation before the end of the Session. It was therefore re-appointed early in the following parliamentary Session, and met twelve times between 2 March and 9 May 1922 (*Report from the Select Committee on Performing Animals, together with the Proceedings of the Committee and Minutes of Evidence*. London: HMSO, 1921 and 1922). The findings of the Select Committee were the basis of the Performing Animals (Regulation) Act, passed in 1925.
2. As part of the Animal Welfare Act (2006), new regulations on the use of animals in circuses may soon be considered, although only 33 wild animals were left in circuses in the United Kingdom in 2007 (Staff. 2007. *Circuses—Help animals leave the circus*. <www.rspca.org.uk> Accessed 17 September, 2007). No animal is known to be performing on any surviving music-hall stage.
3. For example, in 1996 the Born Free Foundation was invited by the Associate Parliamentary Group for Animal Welfare to join the Circus Working Group which published a report on the UK animal circus industry in 1998, *A Report into the Welfare of Circus Animals in England and Wales*. In 2006, the RSPCA joined forces with the Foundation to present a report to MPs outlining the reasons why the welfare of wild animals in the circus cannot be satisfactorily guaranteed due to the nature of circus life: *It's Time Parliament Changed its Act. An Examination of the State of UK Circuses with Wild Animals*.

4. During the debate on the second reading of the Performing Animals (Regulation) Bill, a member cited the RSPCA's annual report for 1919 indicating 1494 members of the Club and its steady growth and influence *(Parliamentary Debates (Commons)* 1923, vol. 161, column 2978).
5. A vice-president of the RSPCA and opponent of vivisection, he was Conservative MP for Epping from 1892 to 1917, when he was created 1st Baron Lambourne.
6. Bertram Mills, as proprietor of the International Circus and Fun Fair at London Olympia, in proposing a toast to the Corporation of the City of London, said some of "the first people in the country" had received letters from opponents, asking them not to visit circuses and opposing the Olympia event, and that Lord Lonsdale had told him the Royal Family said it had also been approached (*World's Fair*, 29 December 1923, front page). Toole Stott (1971, p. 63) suggests that Mills himself was eventually denied a knighthood because "during his last years the anti-animal people were very vocal, and it was this probably which induced the body politic to take the diplomatic line."
7. *Punch*, vol. CLX, 8 June 1921, p. 453; vol. CLX, 29 June 1921, p. 512; vol. CLXIV 13 June 1923, p. 567; vol. CLXVI, 27 February 1924, p. 221. An example of Kenworthy's clumsiness was his question to a Select Committee witness, George Lockhart, elephant trainer and equestrian director of the Tower Circus, Blackpool: "Are you the gentleman who was said to have been killed by an elephant?—No. That was my father" (*Report from the Select Committee on Performing Animals, Together with the Proceedings of the Committee and Minutes of Evidence*. London: HMSO, 1921, paragraph 1308).
8. When the Home Secretary said he could not stop a Christmas circus performance at London's Olympia of 70 lions in a cage with an unarmed man, Kenworthy asked: "Is the Home Secretary not aware that the lion is the national symbol, and that this is bringing the lion into contempt?" (*World's Fair*, 19 December 1925, p. 9).
9. Lord Lonsdale had been criticized by opponents of animal performance because he was at the same time a vice-president of the RSPCA and president of Bertram Mills's International Circus at London Olympia. (He was also president of the Shikar Club. See Callum McKenzie 2000.) He declared that he had accepted the circus presidency because he knew how every trick could be taught, and by accepting it he could ensure that no trick was allowed where cruelty was entailed or necessary. "I made it a condition that all what we call 'acrobatic tricks' were eliminated from the show, and, having seen the turns and things that were done, I am quite convinced that all the tricks performed are done without any cruelty to animals at all" (*World's Fair*, 20 January 1923, p. 16). He was later hissed and booed at an annual meeting of Our Dumb Friends' League because of his attitude to the Rodeo, and Lady Lumb unsuccessfully opposed his re-election as its president on the grounds that she saw no cruelty in it (*World's Fair*, 12 July 1924, p. 4). Soon after, he stated to the press that he was instrumental in inducing the promoters both to cease the public displays of steer roping at the Rodeo, and to allow the RSPCA full access (*World's Fair*, 19 July 1924, p. 10).
10. The following month, Margaret Bradish complained to the *Performer* that she had been quoted by Molloy in his parliamentary speech from a private letter addressed to her by Miss Jessy Wade, secretary of the PADL: "May I ask the trainers and their friends: by what right they read and showed to others my private letter? How they obtained possession of it? Its publication has thrown a curious light on the trainers' methods of obtaining information!" A trainer, Harry Rochez, replied that the VAF said the letter was mistakenly put in with another "usual" one from the PADL to a theatrical manager. He referred to the "lady who is good at shouting" (a Mrs Massingham) as the paid organizer of the PADL, and Miss Bradish as formerly its secretary ("Invited protests against animal acts," *Performer*, 5 April 1923, p. 10; *Performer*, 11 April 1923, p. 14).
11. Mitchell was an anatomist who was appointed secretary of the Zoological Society of London in 1903. He was elected a Fellow of the Royal Society in 1906, and later served between 1923 and 1927 as president of the Society for the Preservation of the Fauna of the Empire. He worked also for the *Times* between 1919 and 1932, writing leaders and scientific articles.
12. Soon after, the sea lion trainer Joseph Woodward drafted and sent a letter to Cockerill, signed by himself (on behalf of the VAF), Frank Glenister (Entertainment Protection Association), R. H. Gillespie (Moss Empires Ltd), Sir Oswald Stoll (Stoll's Circuit), Walter Payne (Syndicate Halls), Percy B. Broadhead (Provincial Entertainments Proprietors' and Managers' Association), John Swallow and Frank Ginnett (Metropolitan Circus Industry), James Sanger and George Harrop (Provincial Circus Industry), Bertram W. Mills (Olympia Circus, London), E. H. Bostock and G. Tyrwitt Drake (Provincial Zoological and Menagerie Industry), W. Savage (Showmen's Guild), and Albert Voyce (VAF). The letter congratulated Cockerill and expressed support for the legislation. In reply, Cockerill said he recognized the spirit of cooperation shown after his taking charge of the unsettled 1924 Bill, and he understood the resentment at indiscriminate attacks that had led to earlier dissension over the Bill. He thought the Bill good for the industry, and there was no excuse now for interrupting performing animal acts because safeguards were in place (*World's Fair*, 4 July 1925, p. 9).

References

Bensusan, S. L. 1913. The case of performing animals (reprinted from *Black and White*, 1899). In *The Under Dog. A Series of Papers by Various Authors on the Wrongs Suffered by Animals at the Hand of Man*, 118–121, ed. S. Trist. London: The Animals' Guardian Office.

Bondeson, J. 2006. *The Cat Orchestra and the Elephant Butler. The Strange History of Amazing Animals.* Stroud: Tempus.

Kean, H. 1998. *Animal Rights: Political and Social Change in Britain since 1800.* London: Reaktion.

Kenworthy, J. M. 1933. *Sailors, Statesmen—and Others. An Autobiography, etc.* London: Rich & Cowan.

Limon, D. and McKay, W. R. eds. 1997. *Erskine May's Treatise on the Law, Privileges, Proceedings and Usage of Parliament.* 22nd edn. London: Butterworth.

London, J. 1917. *Michael, Brother of Jerry.* London: Mills and Boon.

McKenzie, C. 2000. The British big-game hunting tradition, masculinity and fraternalism with particular reference to the "The Shikar Club." *The Sports Historian* 20(1): 70–96.

Ryder, R. D. 1983. *Victims of Science. The Use of Animals in Research.* 2nd edn. London: National Anti-Vivisection Society.

Toole Stott, R. 1971. *Circus and Allied Arts. A World Bibliography. Vol. 4.* Derby: Harpur and Sons.

Turner, E. S. 1992. *All Heaven in a Rage.* 2nd edn. Fontwell, Sussex: Centaur Press.

Wilson, D. A. H. 2001. Sea lions, greasepaint and the U-boat threat: Admiralty scientists turn to the music hall in 1916. *Notes and Records of The Royal Society* 55(3): 425–455.

Wilson, D. A. H. In press a. Racial prejudice and the performing animals controversy in early twentieth-century Britain. *Society & Animals*.

Wilson, D. A. H. In press b. "Crank legislators", "faddists" and professionals' defence of animal performance in 1920s Britain. *Early Popular Visual Culture*.

The State of the Animals IV 2007

Edited by Deborah J. Salem and Andrew N. Rowan

In the fourth volume in the State of the Animals series, a stellar lineup of researchers, scholars, and leaders in the field explore current and emerging issues in animal protection.

JUST PUBLISHED

The more people know about animals, the more they will care. This book sends a powerful message promoting thoughtfulness and compassion over cruelty and ignorance. It provides a breath of fresh air in a world stagnating in animal cruelty and abuse.

— Nicholas Dodman, BVMS, MRCVS, DACVB
Director, Animal Behavior Clinic
Professor, Department of Clinical Sciences
Tufts Cummings School of Veterinary Medicine
Author, *If Only They Could Speak*

The State of the Animals IV: 2007 *continues the tradition established by the previous volumes of providing timely and scholarly review of many core issues in animal protection. It also provides insightful material on bringing action for animals into the 21st century by applying new tactics grounded in use of information technology and economics. A must-read for scholars, students, and activists alike.*

— Randall Lockwood, Ph.D.
Senior Vice President
Anticruelty Initiatives and Legislative Services
American Society for the Prevention of Cruelty to Animals

ISBN 0-9748400-9-2 • Paper • 8½ x 11 • $29.95, $3.00 s/h U.S. orders

Please direct orders to Humane Society Press.

2100 L Street, NW, Washington, DC 20037
202-452-1100; dsalem@humanesociety.org
Online ordering at *humanesocietypress.org*

THE HUMANE SOCIETY OF THE UNITED STATES

Domestic Dogs as Facilitators in Social Interaction: An Evaluation of Helping and Courtship Behaviors

Nicolas Guéguen* and Serge Ciccotti[†]

Université de Bretagne-Sud, Lorient, France
[†] *Université de Bretagne-Sud, Vannes, France*

Address for correspondence:
Dr Nicolas Guéguen,
Université de Bretagne-Sud,
Laboratoire Lestic, UFR
LSHS, 4 rue Jean Zay,
BP 92116,
56321 Lorient Cedex,
France.
E-mail:
nicolas.gueguen@univ-ubs.fr

ABSTRACT Previous studies have suggested that dogs facilitate social interaction between humans. Furthermore, the nature of social interaction is limited to nonverbal behavior such as smiling or gazing or to commonplace conversations. Four studies were carried out in field settings in order to explore if dogs can facilitate closer relationships. In the first experiment, a male confederate (accompanied or not by a dog) solicited people for money in the street. The second experiment was the same except that a female confederate was used. In a third experiment, a male confederate (with or without a dog) accidentally dropped some coins on the ground, to see if people would help him pick them up. In the fourth experiment, a male confederate (with or without a dog) solicited young women in the street for their phone numbers. Results show that the presence of the dog was associated with a higher rate of helping behavior (experiments 1, 2, 3) and higher compliance with the request of the confederate (experiment 4). The influence of a domestic dog as a facilitator to create affiliation and relations in social interaction is discussed.

Keywords: domestic dogs, helping behavior, social interaction

 Various researchers have studied the role of animals in human social interaction. In 1975, Mugford and M'Comisky found that elderly individuals who were provided with a caged budgerigar engaged in more social interaction than those given a houseplant or nothing at all. Similarly, Hunt, Hart and Gomulkiewicz (1992) explored the role of small animals, such as rabbits or turtles, in social interaction between strangers in a park. They found that in a community setting, without special effort or obvious need on the part of a young female confederate, the presence of small animals led unfamiliar children and adults to approach her more often and engage more favorably in conversation with her.

The role of domestic dogs in human interaction has been dealt with in several studies. Some studies suggest that the mere presence of a dog reduces aggression and agitation and promotes social behavior in people with dementia (Filan and Llewellyn-Jones 2006). Bernstein, Friedman and Malaspina (2000) found that during visits of rescue-sheltered dogs, social interaction between residents of nursing homes increased.

Acknowledgement between strangers is also influenced by the mere presence of a domestic dog. Individuals walking through a park with a dog were more likely to receive social acknowledgement from strangers than when they were walking alone (Messent 1984). McNicholas and Collis (2000) found in two observational studies that, within a range of normal daily activities in which a dog could be included and not confined to conventional dog walking areas, the presence of a dog increased the frequency of social interaction, especially interaction between strangers. These authors also found, by varying the apparel of the male confederate (accompanied or not by the dog), an increase in interaction when the confederate was smartly dressed. Furthermore, it was found that, irrespective of the person's style of clothing, the greatest effect was between the dog-present condition and the no-dog control condition.

In shopping malls and on school playgrounds, Mader, Hart and Bergin (1989) recorded the behaviors of passers-by in response to children in wheelchairs. In both settings, it was found that social acknowledgements such as friendly glances, smiles, and conversations were substantially more frequent when the service dog was present. The same effect was found with adults in wheelchairs (Hart, Hart and Bergin 1987; Eddy, Hart and Boltz 1988). For adults without any handicap, the same effect has been found. Fridlund and MacDonald (1998) tested the effect of a Golden Retriever puppy with a human companion on the approaches of passers-by. During their experiment, the puppy aged from 10 weeks to 33 weeks. It was found that approaches were most numerous when the puppy was youngest, and females approached more often than males during the first half of sampling, but approached a similar number of times to males during the second half. In a recent study carried out by Wells (2004) in a field setting, 1800 male and female pedestrians approaching a female experimenter were observed according to the presence of three dogs (Labrador Retriever pup, Labrador adult, Rottweiler adult), two neutral stimuli (teddy bear, potted plant), or without any accompaniment. The acknowledgement of the pedestrians were unobtrusively observed and coded with a grid of five levels of social interest (ignore, completely overlook the experimenter, look at the experimenter, smile at the experimenter, or talk to the experimenter). It was found that more people ignored the experimenter when accompanied by no stimulus, neutral stimuli, and the adult Rottweiler, whereas more pedestrians smiled at, or talked to, the experimenter when accompanied by the Labrador pup or adult. When the length of conversations was measured, it was found that passers-by engaged in longer conversations when she was accompanied by the Labrador pup than when she was with the Labrador adult.

Overall, the studies discussed above found that animals, and particularly domestic dogs, are associated with increased social interaction between humans. However, these studies focused only on brief social interaction. It would be interesting to explore the effect of domestic dogs on more elaborate or closer relations between strangers. To that end, we examined the role of domestic dogs in human helping and courtship behaviors in four experiments—a male confederate (a female confederate in one study) accompanied or not by a dog, solicited pedestrians for their help, and in one study solicited women for their phone numbers. Because previous studies have shown that domestic dogs enhance social interaction, we hypothesized that the presence of a dog would lead to greater compliance with a person's request, compared with when no dog was present.

Experiment 1

Methods

Participants: Eighty men and 80 women (age range approximately 25–60 years old) were chosen at random in the street. Eighty were randomly assigned to the experimental group (40 men and 40 women) and 80 were assigned to the control group (40 men and 40 women).

The confederate was a 22-year-old male of medium height (1.75 m) and weight (71.2 kg). He was neatly dressed and in a conventional way for a person of his age (jeans/sneakers/T-shirt).

This dog used in this study was a mongrel of medium height (42 cm) and weight (11.4 kg), with a black, mid-length coat. In a previous evaluation conducted in the street with 47 men and women, this dog was evaluated as kind, dynamic, and pleasant.

Procedure: The experiment took place in a public mall, during sunny, spring days in 2006. The confederate was instructed he had to approach the first man or woman, aged from 25 to 60 years old, who walked alone in the pedestrian zone. When the confederate was ready to solicit people, he was instructed to count from one to five pedestrians and then approach the first appropriate passer-by. With this method, it was not possible for the confederate to select a participant according to his subjective appreciation. If the first passer-by was a child, a teenager, an elderly man/woman, or a person in a group, the confederate had to ignore him/her and wait until an appropriate person passed by. After finishing with a participant, the confederate was instructed to count the next five pedestrians and then approach the next appropriate person.

In the experimental condition, the confederate kept a dog on a lead. In the no-dog control condition, the confederate was not accompanied by the dog, but was instructed to interact in the same way with the participants.

The confederate, who owned a dog but not the dog used in this experiment, was trained before the experiment was conducted. He was instructed to approach 10 people, accompanied or not by the dog, and his conversation was recorded using a digital recorder (Roland Edirol R1) which was placed in the inside pocket of his jacket. A very discrete microphone was placed on the lapel of his jacket (although small, the microphone was highly sensitive, enabling high-quality recordings to be made). Five judges (two males, 22 years old each and three females, 20–22 years old) listened to the recordings and were instructed to evaluate if the confederate was accompanied by a dog or not when soliciting the pedestrian. It was found that the discriminatory capabilities of the judges were not statistically different to what would be expected by chance.

In the experiment, the confederate approached each person and said, politely, "Sorry Madam/Sir would you have some money so that I can catch the bus, please ?" In the case of a positive answer, the confederate waited for the participant to give him some money. He estimated the amount given and then gave it back to the participant. The person was then debriefed about the study.

The dependant variables used in this experiment were the number of participants who complied with the confederate's request and the amount of money given by participants who had agreed to make a donation. The first dependant variable we used was in a 2×2 chi-square test, in order to test the relationship between the two dichotomous variables (experimental conditions: dog/no dog and compliance with the request: comply/not comply). The second dependant variable was a continuous variable, so we used a *t*-test in order to evaluate if the mean donation in the dog condition was statistically different from the mean donation in the no-dog condition.

Results and Discussion

On all measures employed in this study, no differences were found between male and female participants according to the experimental conditions, so data were aggregated. In the no-dog control condition, 11.3% (9/80) of the participants solicited complied with the confederate's request, compared with 35% (28/80) in the experimental condition. This difference was significant ($\chi^2_{(1)} = 15.26, n = 160, p < 0.001, r = 0.30$). The presence of the dog was associated with higher compliance with the confederate's request.

When we considered the mean amount of money donated by the people having accepted the request in each of the experimental groups, we found that participants gave an average of €0.63 (US$0.26) in the no-dog condition versus €0.87 (US$0.31) in the dog condition. This difference was significant ($t_{(35)} = 2.09, p < 0.05$, two-tailed, $d = 0.71$). Thus, the presence of the dog was associated with greater levels of generosity from the participants who agreed to the confederate's request.

Two pro-social effects were associated with the presence of the dog in this experiment. We found that a higher number of pedestrians helped the confederate when he was accompanied by a dog, and the pedestrians who complied with the request were more generous with their donations when the confederate was accompanied by the dog than when he was not. These effects show that altruism is strongly connected with the presence of a domestic dog. Previous research found that pedestrians' acknowledgement of a stranger increased when he/she was accompanied by a dog (Wells 2004) and that more people engaged in conversations when a dog was present. In this experiment, we found that more intimate social behaviors are also affected by the presence of a dog. Such effects suggest that the dog has a high ability to affect social interaction between humans.

In order to study this effect across the sexes, a second experiment was carried out, this time using a female confederate. Helping behavior has been found to be affected by the gender of the person seeking help (Bickman 1971; Juni and Roth 1981; Basow and Crawley 1982; Fiala et al. 1999)—women, more than men, receive help when soliciting it in the street.

Experiment 2

Methods

Participants: One hundred men and 100 women (age range approximately 25–60 years old), were chosen at random in the street. One hundred people were randomly assigned to the experimental group (50 men and 50 women) and 100 to the control group (50 men and 50 women).

The confederate was a 21-year-old female of medium height (1.66 m) and weight (59 kg). She was dressed neatly and in a conventional way for people of that age (jeans/sneakers/pullover/jacket).

The dog used in the experimental condition was the same as in the previous experiment, and the verbal solicitation was strictly the same as in the first experiment.

Procedure: The experiment took place in a public mall, during sunny, but cold, winter days in December 2007. The confederate approached appropriate passers-by using the same methodology as in the first experiment. A training period was used in order to prevent any difference in the behavior of the confederate in the two experimental conditions. As in the previous experiment, verbal behavior was recorded and evaluated by judges—they were not able to discriminate between the two experimental conditions. As before, the response of the participant to the confederate's request and the amount of money given were the two dependant variables. Data were analyzed using the same statistical procedures as described in experiment 1.

Results and Discussion
No differences were found between male and female participants according to the experimental conditions, but in both conditions men complied more favorably with the request (52%) than women (25%). No interaction effect was found between the participants' gender and the experimental conditions, so data were aggregated. In the no-dog control condition, 26% (26/100) of the participants solicited complied with the confederate's request, compared with 51% (51/100) in the experimental condition. This difference was significant ($\chi^2_{(1)} = 13.20$, $n = 200$, $p < 0.001$, $r = 0.26$). The presence of the dog was associated with higher compliance with the confederate's request. When we considered the mean amount of money donated by the people having accepted the request in each of the experimental groups, we found that participants gave a mean of €0.54 (US$0.31) in the no-dog condition versus €0.80 (US$ 0.47) in the dog condition. This difference was significant ($t_{(75)} = 2.55$, $p = 0.02$, two-tailed, $d = 0.59$) and confirms that the presence of the dog was associated with greater levels of generosity of the participants who agreed to the confederate's request.

In the last two experiments, an explicit verbal request addressed to people in the street was found to be more favorably accepted when the confederate was accompanied by a dog. In order to compare the interaction effect between the gender of the confederate and experimental condition on compliance with the confederate's request, a 2 (dog/no-dog condition) × 2 (male/female confederate) × 2 (compliance/no compliance) log linear analysis was performed. Participants were found to comply more favorably ($G^2_{(1, 360)} = 9.99$, $p < 0.005$) with the female confederate's request (38.5%) than with the male confederate's request (23.1%). These results are in accordance with previous research on helping behavior which showed that females elicited greater compliance with a request than males (Bickman 1971; Juni and Roth 1981; Basow and Crawley 1982; Fiala et al. 1999). Furthermore, an interaction effect was also found between the gender of the confederate and the experimental conditions ($G^2_{(4, 360)} = 36.45$, $p < 0.001$) and revealed that the effect of the dog was greater with a male confederate (the gain of compliance increased more than threefold) than with a female confederate (the gain of compliance increased less than twofold). Therefore, the gain of compliance exerted by the mere presence of the dog seems to be greater with the male than with a female.

With the mean amounts of money donated by the people having accepted the request, a 2 (dog/no-dog condition) × 2 (male/female confederate) ANOVA (Analysis of Variance) was performed. A main effect of the experimental condition was found ($F_{(1, 110)} = 8.09$, $p < 0.001$, $Eta^2 = 0.13$) but no main effect of the gender of the confederate was found ($F_{(1, 110)} = 0.82$, ns, $Eta^2 = 0.04$), and no interaction between the two main factors was found ($F_{(1, 110)} = 0.02$, ns, $Eta^2 = 0.00$). So it seems that gender had no effect on the amount of the donation and does not interact with the presence versus absence of the dog.

This method to test altruism is a traditional one used in social psychology to explore the effects of some variables on helping behavior (Bierhoff 2002). However, the methodology used in research which focuses on helping behavior not only uses explicit request, but also implicit request. An implicit helping request is a request that is not directly addressed by one person to another: it is a situation where someone decides to help a person spontaneously. For example, a situation where someone drops coins or papers on the ground without soliciting somebody to help him/her to pick them up is typically one where spontaneous helping behavior could occur: persons around, without solicitation, could decide themselves to help the person. This technique is frequently used in studies on helping behavior, and many scientists consider that it is a better method for studying the determinants of altruism because people who provide their help are not solicited and are free to choose—they could be considered Good Samaritans (Bierhoff 2002). Our third experiment, carried out again in a field setting (bus shelter), tested the effect of the presence of a domestic dog on the spontaneous helping behavior of strangers when something happens to a person but there is no solicitation for help.

Experiment 3
Methods
Participants: Forty men and 40 women (age range approximately 25–60 years old), who were sitting down and waiting in a bus shelter, were chosen at random.

The same male confederate who was used in the first experiment was used in this study. He was dressed in the same way as before. In the experimental condition, the confederate was accompanied by the same dog as before. Again, the dog was kept on a lead.

Procedure: An observer (a woman) in front of a bus shelter (30 meters away) waited until a man or a woman (approximately aged from 25 to 60 years) entered the bus shelter. She then phoned the confederate, who was waiting with the dog in a car 50 meters behind the bus shelter. The observer described the person who had just arrived and told the confederate to go to the bus shelter with or without the dog (this was randomized). The confederate had to enter the bus shelter, look at his watch, and then pretend to be engrossed in the bus timetable. He was instructed not to look at the other people during the time he was in the bus shelter. After 30 seconds, the confederate had to leave the bus shelter. While leaving, he had to put his hand in his pocket, take out a handkerchief, and "accidentally" drop some coins on the ground. The confederate was then instructed to wait two seconds before bending down to pick up the coins. Again, in both conditions (experimental and control), the confederate was instructed not to look at the participants when acting. If anyone helped him pick up the coins, the confederate was instructed to thank him/her and leave the bus shelter. If no help was provided by anyone, the confederate was instructed to pick up all the coins and leave the bus shelter. Overall, the confederate made 80 separate trips to the bus shelter.

The dependant variable used in this experiment was the number of participants who helped the confederate. To analyze our data we used a 2 × 2 chi-square test, in order to test the relationship between the two dichotomous variables (experimental conditions: dog/no dog and helping behavior: help/no help).

Results and Discussion
Between all measures employed in this study, no differences were found between male and female participants according to the experimental conditions. Again, data were aggregated. In the experimental condition, in which the confederate was accompanied by a dog, 87.5% of the participants (35/40) helped the confederate, compared with 57.5% (23/40) in the no-dog control condition. This difference is significant ($\chi^2_{(1)} = 9.03$, $n = 80$, $p < 0.005$, $r = 0.34$).

For the third time, with a different methodology, it was found that the presence of a dog elicited greater helping behavior. In this experiment, the effect of the dog is more interesting because the altruistic behavior tested was spontaneous and no previous verbal or non-verbal interaction occurred between the confederate and the participants. The mere presence of the dog seems to have been sufficient to elicit greater helping behavior toward the confederate. These data suggest that the domestic dog is really a facilitator of social interaction that not only encourages social behaviors such as smiles or greetings (Wells 2004), but also a different range of social behaviors such as helping. In order to explore this effect further, a fourth experiment was conducted that tested the effect of the presence of a dog on highly intimate interactions and solicitation. In this new experiment, the effect of the presence of a domestic dog on a young man's courtship solicitation toward young women was tested.

Experiment 4
Methods
Participants: The participants were 240 young women (ranging in age from approximately 18–25 years) chosen at random while walking alone in a pedestrian zone in the same city where the two first experiments were conducted.

In this experiment, a 20-year-old, male confederate was used. He was selected by a group of three women, on the basis of a high physical attractiveness score, from a group of three male volunteers. An attractive man was used because pre-test evaluation showed that it was generally difficult to obtain phone numbers from young women in the street. This avoided creating conditions where the ceiling effect of compliance was low. The confederate wore the same style of clothing as in experiments 1 and 3.

The same dog that was used in the previous studies was used.

Procedure: The experiment was carried out on sunny days in July 2007. In this experiment, the participants were selected following a random assignment in which the confederate was instructed to approach the first young woman in the relevant age group (18 to 25 years) who was walking alone in the pedestrian zone where the experiment was being carried out. As in the previous experiments, when soliciting the young women the confederate kept his dog on a lead in the experimental condition, whereas he was not accompanied by his dog in the no-dog control condition. The same verbal solicitation was made by the confederate in both the control and the experimental conditions:

> *"Hello. My name's Antoine. I just want to say that I think you're really pretty. I have to go to work this afternoon, but I was wondering if you would give me your phone number. I'll phone you later and we can have a drink together someplace."*

The confederate, who owned a dog, but not the dog used in this experiment, was trained before beginning the experiment. He was instructed to approach 10 participants, accompanied or not by the dog, and his conversation was recorded by a digital recorder (Roland Edirol R1), as in experiment 1. Five judges (2 males, 22 years old and 3 females, 20–22 years old) listened to the recordings and were instructed to evaluate if the confederate was accompanied by a dog or not when soliciting the pedestrian. It was found that the discriminatory capabilities of the judges were not statistically different to what would be expected by chance.

In the experimental situation, after making his request, the confederate was instructed to wait 10 seconds and to gaze and smile at the participant. If the participant accepted the confederate's solicitation, the confederate debriefed her about the study. An information sheet was then given to her and she was asked to provide some details for the experiment (name, age, address, phone number). The information sheet contained details of the project, the laboratory's web site address, and the personal phone number of the director of the laboratory. To date (the fourth experiment was conducted in July 2007), no participant has phoned to obtain further information about the research. This method of debriefing was used in our experiment because firstly, this was the recommendation of the ethical committee of our laboratory, and, secondly, in previous experiments where the same request was solicited (Guéguen 2007a), a more in-depth debriefing had been performed (contacting the young women two days later), revealing that none of the participants had been insulted or troubled by the experiment. Indeed, most mentioned that it had been amusing to participate in such an experiment and that they had good and pleasant memories of it.

After debriefing each participant in our study, the confederate would say, *"Thanks for your participation and I'm sorry that I've taken up your time. Perhaps we could meet another time. Bye!"* If the participant had refused to give her number, the confederate was instructed to say, *"Too bad. It's not my day. Have a nice afternoon!"* and wait for another participant.

The dependant variable used in this experiment was the number of women who complied with the confederate's request by giving him her phone number. To analyze the data we used a 2 × 2 chi-square test to test the relationship between two dichotomous variables (experimental conditions: dog/no dog and compliance to the request: comply/not comply).

Results and Discussion

The dependent variable in this experiment was the number of participants who agreed to the courtship request. The results showed that 28.3% (34/120) of the women approached complied with the request when the confederate was with the dog, compared with 9.2% (11/120) of women when the confederate was not with the dog. This difference is significant ($\chi^2_{(1)}$ = 14.47, $n = 240$, $p < 0.001$; $r = 0.25$).

For the fourth time, the presence of a domestic dog was found to have a positive impact on social interaction. Furthermore, in this experiment, the request of the male confederate had a high level of social intimacy. Despite this intimate request, the effect of the presence of the dog still remained, proving that a dog is a powerful facilitator of social interaction.

In a previous study on courtship behavior using the same methodology (Guéguen 2007a), it was found that tactile contact (young women were touched lightly by a male confederate when he asked for their phone number) elicited greater compliance with a courtship solicitation, but the rate of compliance (19.2%) is lower than the rate of compliance found in the current study (28.3%). However, the control groups in the two studies had similar rates of compliance: 10% in the no-touch control group (Guéguen 2007a) and 9.2% in the no-dog control group (current study). The effect size of the effect of touch ($r = 0.13$) and the effect size found here ($r = 0.25$) confirms the more powerful effect of the domestic dog. Given that these two experiments were similar methodologically (same confederate, same setting and period of testing and same request), the difference found suggests the higher efficiency of the domestic dog in social interaction.

General Discussion

The results of our four experiments show that requests addressed to pedestrians by a stranger accompanied by a dog are more favorably received than when the person is not accompanied by a dog. This agrees with previous research which found that dogs can enhance social interactions between humans and that a stranger accompanied by a dog receives more acknowledgements. In these earlier studies, the interactions were limited to greetings. However, in our experiments, we used more elaborate and intimate social behaviors, and the results highlight that a large spectrum of social behaviors is influenced by the presence of dogs. Overall, our results confirm the social lubrication effect of dogs (Wells 2004).

Our purpose was only to study the effect of the domestic dog on various social behaviors associated with closeness and intimate social interactions. In addition, these behavioral effects have to be explained theoretically. A host of previous research has found that helping behavior toward a stranger is influenced by factors associated with the person, such as physical attractiveness (Harrel 1978; Nadler, Shapira and Ben-Itzhar 1982), apparel (Chierco, Rosa and Kayson 1982; Sinha and Jain 1986), and even the model of the person's car (Solomon and Herman 1977). Our results confirm that a dog is another factor that enhances the attractiveness of a person and helps elicit greater helping behavior toward him/her. Furthermore, some differences exist between the latter studies on helping behavior and the experiments conducted here. Indeed, in the earlier research, the apparel of the experimenter or the value of his/her car was associated with different levels of social status which were associated with his/her appearance. We think that the presence of a dog has no impact on such evaluations but could influence how a person is perceived. In our experiments, we think that the confederate accompanied by his dog was differently perceived on personal attributes than when he was not accompanied by the dog. Perhaps the presence of the dog led pedestrians to evaluate the confederate to be more kind, thoughtful, or sensitive, possibly because people who love animals, particularly dogs, are thought to have these attributes. In return, this evaluation of personal attributes may have led the pedestrians to be more agreeable and helpful. The positive effect of the young women toward the confederate in our fourth experiment is congruent with this explanation, given that positive social attributes of men are generally associated with greater attractiveness in a courtship context (Guéguen 2007b). Of course, this explanation in terms of the activation of positive social attributes associated with the confederate accompanied by a domestic dog remains speculative, given that the evaluation of such attributes was not performed. It would be interesting in future studies to evaluate the link between the presence of the dog and the personal attributes associated with his/her owner.

Of course, the results from our four experiments cannot be generalized to every dog. In our experiments, the same, black dog was used every time. Wells and Hepper (1992) found that people preferred blonde to black-coated dogs. Because of this, we hypothesize that greater helping behavior and compliance with a courtship request could be obtained with a yellow/light-coated dog. Also, Wells (2004) found that a female experimenter accompanied by a Labrador received more acknowledgements from pedestrians than when she was accompanied by a Rottweiler. Our experiments, however, were conducted using a mongrel, not a pedigree dog. It will be necessary in future studies to further test the influence of breed and color of dog on social interaction, particularly when trying to solicit helping behavior from strangers.

In our first and second experiments, we found no difference in helping behaviors in male and female pedestrians, which could be explained by the fact that we used an adult domestic dog. Fridlund and MacDonald (1998) found that a Golden Retriever puppy elicited a greater number of female pedestrian approaches than male pedestrian approaches, suggesting a human female preference for canine juvenescence. Therefore, regarding our fourth experiment, where a positive effect of the dog was found on eliciting positive responses to a courtship request addressed to young women in the street, we hypothesize that had we used a puppy, greater compliance with the request would have been achieved.

There are some other methodological problems in the experiments we conducted. In experiments 3 and 4, only young male confederates were used and so the results cannot be generalized for both sexes. We found in experiments 1 and 2 an interaction effect between the gender of the confederate and the presence versus absence of the dog on compliance with a request. With an explicit helping solicitation, the gain of compliance exerted by the mere presence of the dog was greater with the male confederate than with the female confederate. Such results need to be confirmed, but this gender effect is interesting—it shows that perhaps people react differently to a stranger accompanied by a dog according to the gender of the person. Thus, it will be interesting for further studies to evaluate reactions and representations associated with the presence of a dog according to its owner's gender.

The confederates in our studies were young and so the data are not generalizable to all age groups. Also, the results cannot be generalized to every culture, given that these experiments were conducted in France only, a country where dogs are very popular. Additional data are now necessary to explore the generality of the enhancing effect of domestic dogs on social interaction, to evaluate the factors associated with greater or lower efficiency of this effect, and to evaluate the theoretical explanations for it.

A possible confederate bias might have been present in our experiments. While the confederate was instructed to behave identically when soliciting participants, and we established in the first, second, and fourth experiments that there were no differences in the two conditions (with or without dog) according to the verbal behavior of the confederate, variation in nonverbal behavior may have occurred. Of course this bias might be present in all of the previous studies cited in this paper, too, but it is important to evaluate this effect in future studies by examining videotapes of the interactions, by increasing the number of confederates, and by training them more fully.

In summary, congruent with previous studies, our data on helping behavior and request solicitation confirm the positive role of domestic dogs in social interaction between strangers.

Acknowledgements
The authors thank two anonymous reviewers for their help to improve the quality of this paper

References
Basow, S. and Crawley, D. 1982. Helping behavior: Effects of sex and sex-typing. *Social Behavior and Personality* 10: 69–72.

Bernstein, P. L., Friedmann, E. and Malaspina, A. 2000. Animal-assisted therapy enhances resident social interaction and initiation in long-term care facilities. *Anthrozoös* 13: 213–223.

Bickman, L. 1971. The effect of social status on the honesty of others. *The Journal of Social Psychology* 85: 87–92.

Bierhoff, H.-W. 2002. *Prosocial Behaviour.* Hove: Psychology Press.

Chierco, S., Rosa, C. and Kayson, W. 1982. Effects of location, appearance, and monetary value on altruistic behavior. *Psychological Reports* 51: 199–202.

Eddy, T. J., Hart, L. A. and Boltz, R. P. 1988. The effects of service dogs on social acknowledgments of people in wheelchairs. *Journal of Psychology* 122: 39–45.

Fiala, S., Giuliano, T., Remlinger, N. and Braithwaite, L. 1999. Lending a helping hand: Effects of gender stereotypes and gender on likelihood of helping. *Journal of Applied Social Psychology* 29: 2164–2176.

Filan, S. L. and Llewellyn-Jones, R. H. 2006. Animal-assisted therapy for dementia: A review of the literature. *International Psychogeriatrics* 18: 597–611.

Fridlund, A. J. and MacDonald, M. 1998. Approaches to Goldie: A field study of human response to canine juvenescence. *Anthrozoös* 11: 95–100.

Guéguen, N. 2007a. The effect of a man's touch on woman's compliance to a request in a courtship context. *Social Influence* 22: 81–97.

Guéguen, N. 2007b. *100 Petites Expériences de Psychologie de la Séduction. Pour Mieux Comprendre tous nos Comportements Amoureux.* Paris: Dunod.

Harrel, A. 1978. Physical attractiveness, self-disclosure and helping behavior. *The Journal of Social Psychology* 104: 15–17.

Hart, L. A., Hart, B. L. and Bergin, B. 1987. Socializing effects of service dogs for people with disabilities. *Anthrozoös* 1: 41–44.

Hunt, S. J., Hart, L. A. and Gomulkiewicz, R. 1992. Role of small animals in social interactions between strangers. *Journal of Social Psychology* 132: 245–256.

Juni, S. and Roth, M. 1981. Sexism and handicapism in interpersonal helping. *The Journal of Social Psychology* 115: 175–181.

Mader, B., Hart, L. A. and Bergin, B. 1989. Social acknowledgments for children with disabilities: effects of service dogs. *Child Development* 60: 1529–1534.

McNicholas, J. and Collis, G. M. 2000. Dogs as catalysts for social interactions: Robustness of the effect. *British Journal of Psychology* 91: 61–70.

Messent, P. R. 1984. Correlates and effects of pet ownership. In *The Pet Connection: Its Influence on Our Health and Quality of Life*, 331–340, ed. R. K. Anderson, B. L. Hart and L. A. Hart. Minneapolis, MN: Center to Study Human–Animal Relationships and Environments, University of Minnesota.

Mugford, R. A. and M'Comisky, J. G. 1975. Some recent work on the psychotherapeutic value of caged birds with old people. In *Pets, Animals and Society*, 54–65, ed. R. S. Anderson. London: Bailliere Tindall.

Nadler, A., Shapira, R. and Ben-Itzhar, S. 1982. Good looks may help: Effects of helper's physical attractiveness and sex of helper on males' and females' help seeking behavior. *Journal of Personality and Social Psychology* 41: 90–99.

Sinha, A. and Jain, A. 1986. The effects of benefactor and beneficiary characteristics on helping behavior. *The Journal of Social Psychology* 126: 361–368.

Solomon, H. and Herman, L. 1977. Status symbols and prosocial behavior: the effect of the victim's car on helping. *Journal of Psychology* 97: 271–273.

Wells, D. L. 2004. The facilitation of social interactions by domestic dogs. *Anthrozoös* 17: 340–352.

Wells, D. L. and Hepper, P. G. 1992. The behaviour of dogs in a rescue shelter. *Animal Welfare* 1: 171–186.

COMPANION ANIMALS: FROM PRE-HISTORIC TIMES TO WHAT HAPPENED YESTERDAY

Companion Animals in Society
By Stephen L. Zawistowski, Ph.D., CAAB

Since companion animals are an important part of American society, a substantial body of research has been developed to demonstrate how they play a significant role impacting the physical and psychological health of people of all ages.

Companion Animals in Society combines the current available knowledge on companion animal husbandry with an introduction to these issues and how society is currently coping with them in order to provide the most useful resource in the market today.

Some of the textbook's features include:

- Links the social and biological history of companion animals using the most recent data on the taxonomy of cats and dogs based on modern molecular methods

- Traces the biology and industry of companion animals from pre-historic times to what is occurring in today's society

- Provides insight into the organization and interaction between different companion animal organizations and interest areas

About the Author
Stephen L. Zawistowski is EVP and Science Advisor of the ASPCA®, where he currently oversees the ASPCA's National Programs. "Dr. Z." is a certified applied animal behaviorist and a frequent speaker on animal behavior, the history of sheltering, humane education, and the use of statistics in program management.

Now available online at **www.aspca.org/store**

Attitudes and Actions of Pet Caregivers in New Providence, The Bahamas, in the Context of Those of Their American Counterparts

William J. Fielding
The College of The Bahamas, Nassau, The Bahamas

Address for correspondence:
William J. Fielding,
Planning Office, The College
of The Bahamas,
P. O. Box N-4912, Nassau,
New Providence,
The Bahamas.
E-mail: wfielding@cob.edu.bs

ABSTRACT This paper reports the attitudes and actions of 614 Bahamian pet caregivers towards their pets, irrespective of the type of pet kept. The results are discussed in the context of an American study by Pamela Carlisle-Frank and Joshua Frank, published in 2006, which posed similar questions. While Bahamians appeared to interact less with their pets than Americans (e.g., 56.9% of Bahamians took pets on walks compared with 71.5% of Americans), they had some important attitudes towards animals in common with Americans, such as disapproving of declawing of cats (67.9% of Bahamians and 71.1% of Americans) and long-term chaining of dogs (83.0% of Bahamians and 84.9% of Americans), and the recognition of the need to help animals (85.0% of Bahamians and 90.5% of Americans). Some Bahamian respondents voluntarily suggested that white people and black people cared for animals differently. The results suggest that the differences between the two communities, in terms of actions and attitudes towards pets, may explain why some American visitors to The Bahamas think that Bahamians do not care for their pets. These differences are of potentially great economic importance, as most tourists visiting The Bahamas come from the US.

Keywords: attitudes, pet care, pets, The Bahamas, North America

The history of The Bahamas and the United States of America (US) has been intertwined over many years, from the time of the Eleutheran Adventurers through the Loyalists (who left America to settle in The Bahamas, as they supported British rule), US invasion, and through blockade running during the American prohibition (Craton and

Saunders 1992 and 1998). It is generally accepted that "Bahamians have come more and more under the influence of the US" since the Second World War (Strachan 2007). The Bahamas, with a population of about 300,000 (Department of Statistics 2002), hosts over four million US tourists annually (in 2005, more than 4.3 million US tourists visited [US Department of State 2007]), so there is constant interaction between the two communities. Further, many Bahamians are educated in the US (Scholarship and Educational Loan Division 2003) and 28,075 (or about 10% of the population living in The Bahamas) people born in The Bahamas were living in the US in 2000 (US Census Bureau 2001), so it can be expected they acquire American perspectives which may be retained, should they return home. The close ties between the two countries are, perhaps, exemplified by the US embassy being known as "the" embassy.

Despite their geographical proximity, the US and The Bahamas are ranked 8th and 51st, respectively, in the UNDP Human development Index (UNDP 2006). This difference reflects a disparity in their per capita Gross Domestic Product ($39,676 and $17,843, respectively), education levels, and life expectancy (77.5 and 70.2 years, respectively). Consequently, there are major differences, developmental as well as cultural, between the two communities.

It is estimated that tourism in The Bahamas accounts for over 50% of Gross Domestic Product (World Travel and Tourism Council 2007) and so anything which might threaten this industry is a concern. The presence of millions of American tourists in The Bahamas, 80% of whom are Caucasian (Research and Statistics Division 2005), means that the care offered to animals by this Afro-Caribbean community is open to scrutiny by visitors who are culturally and ethically different. When the level of care is not agreeable to visitors or differs from their cultural norms, this can result in the image of the country being tarnished, and some Americans do get negative impressions of The Bahamas from how they perceive animal care. It has been estimated that the negative impressions US tourists acquire as a result of Bahamian pet care may cost the country 10% of tourism revenues (Plumridge and Fielding 2003). While that study indicated that there are probably important differences in the way Bahamians and Americans care for their pets, they have not yet been assessed. It should also be noted that the last major revision to the Dog License Act (in 1942) was as a result of visitors complaining about roaming dogs (Fielding, Mather and Isaacs 2005). Consequently, the importance of differences between pet care with which tourists are familiar (in The Bahamas this means American tourists) and that which is expressed in The Bahamas, cannot be underestimated. It is important to appreciate differences, if any, in attitudes and actions towards pets between these two communities.

The most common pet in New Providence, the island where the capital Nassau is located, is the dog (Fielding and Plumridge 2005a), but other pets such as cats, fish, and birds are also kept. In the case of dogs, the only pet subjected to much research in New Providence, care usually satisfies the basic necessities, which would be the provision of food, water, and shelter (Fielding and Plumridge 2005b), a level considered "essential care" (Shore, Riley and Douglas 2006). In New Providence, most dogs are kept outside and used for protection (Fielding and Plumridge 2005b), which is not conducive to interaction and forming strong bonds between animal and caregiver (Fielding 2007). Elsewhere, it has been shown that "yard" dogs receive different levels of care than "house" dogs, even if both groups of dogs do receive "essential care" (Shore, Riley and Douglas 2006). The low level of interaction between pets and caregivers in the Caribbean has given rise to the concept of "passive ownership" (Alie et al. 2007). However, it should be noted that in the case of dogs, those within the same household may receive different standards of care, depending upon size and type (breed, cross-breed, mongrel) (Fielding and Plumridge 2005b). Therefore, many factors interact to influence the level of care offered and the interaction between caregiver and pet.

Despite attempts by local animal welfare groups, there is limited education on animal welfare in schools. Further, debates which have occurred elsewhere on animal welfare issues, such as "owners" versus "guardians" (the debate in North America is outlined by Carlisle-Frank and Frank [2006]) have not occurred in The Bahamas. Pets, and in particular dogs, are usually considered "owned," an attitude reinforced by the law (which refers to dogs as having owners) and by some local animal welfare groups which promote "responsible pet ownership" (Proud Paws n.d.). Consequently, the benchmark we will use in this study for comparison with American pet caregivers will be all pet keepers, without distinguishing them as "owner," "guardian," or "owner/guardian" (Carlisle-Frank and Frank 2006).

A previous study on pet attachment in New Providence included college students who were not necessarily pet keepers and so also reported perceived rather than actual attachment (Fielding, Mather and Isaacs 2005). Consequently, the current research is the first known study to exclusively look at pet attachment amongst pet keepers, beyond cats and dogs, in the wider public within an Afro-Caribbean community. Given the tourism economy of the country, it was decided to assess attitudes and actions of Bahamian pet caregivers and to discuss them in relation to those of their American pet-keeping counterparts. This would allow for further examination as to why tourists are sometimes upset at what they perceive as a lack of care towards pets (particularly dogs, as they are highly visible) exhibited by Bahamians (Plumridge and Fielding 2003), and whether the differences are action-based or/and attitude-based.

Methods

A survey form devised by Carlisle-Frank and Frank (2006) was used with slight modification. Questions focused on attitudes towards the family pet, treatment of companion animals, and beliefs and perceptions of animals in general. Some questions were amended to account for local parlance and pet care practices. Results from other Bahamian studies on animals (e.g., Fielding and Plumridge 2005b) and the input from a class of social science students at The College of The Bahamas were used to make these changes. Some dog and cat owners dispose of their deceased pets in the trash (Smith et al. 2006), so the method of disposal was included to see if this was linked with attachment to the pet, as others have suggested (Varner and Johnson Varner 1983). Some questions about, for example, celebrating a pet's birthday and having photographs of a pet, caused considerable debate, as many students thought that the questions were irrelevant to Bahamian pet caregivers, an interesting observation in itself. It should be noted that in The Bahamas, to lose a pet generally means that the pet is lost forever, as many pets are routinely allowed to roam unsupervised. Consequently, this question was made more specific to indicate lost forever; this is different from North American parlance, where many "lost" pets are returned. It was also considered that many people might not understand the term "sentient," which was used by Carlisle-Frank and Frank, so this was replaced by two questions designed to obtain the equivalent information. Finally, a question on motor vehicle ownership was included as an indicator for poverty (Johnson et al. 2005). A summary of the questions is given in Table 1.

No study appears to have been done on relinquishment in The Bahamas, but it is known that in the case of unwanted puppies, many are "given" away and few are taken to a shelter or animal welfare group (Fielding 2007). This suggests that "relinquishment" in this context does not necessarily mean surrendering the pet to an animal welfare group, but rather disposing of the pet in some less formalized way.

Table 1. Questions used in the survey.

Number and type of pets
Source of pets
Satisfaction with pets
Do you consider your pet a member of your family?
Do you consider your pet your property?
Do you feel attached to your pet?
Do you identify with your pet?
Where applicable, are your pets spayed/neutered?
In the past two years, how many times, if any, have any of your pets gone missing (lost forever)?
In the past week, have any of your pets had uncontrolled access to the street?
In the cases where registration is required, are your pets registered and/or licensed?
How many pets/companion animals have you had to give up due to moving home, family problems, or other relationship, personal or family problems?
Where applicable, do your pets have identification such as microchips, tags?
Do you permit your pets to live indoors with you/the rest of the family?
If you celebrate birthdays, do you celebrate your pets' birthdays?
If you have a family photo-album, are your pets' photos included?
Are your pets' names included with other family members on Christmas and/or greeting cards? (If you send out cards)
Where applicable, do you take your pets on family walks, outings, drives, picnics, vacations, or day-trips?
How often do you tell your companion animals that you love them per week?
Do you believe that long-term chaining of dogs should not happen?
Do you believe that spay/neutering should be done to stop overpopulation and suffering of animals?
Do you believe that pets should not live long-term in cages?
Do you believe that viewing pets as possessions is wrong?
Do you believe that we should not make a deal out of protecting pets?
Do you believe that we should help animals because they are dependent on humans/are helpless?
Do you believe that animals have feelings with needs/interests of their own?
Do you believe that animals can feel pain?
Does your household own a motor vehicle?
Do you consider yourself a good caregiver to your pets?
The single main reason for having pets.
Method of disposal of your dead pets.
Gender
Age group

Questions adapted from questionnaire used in Carlisle-Frank and Frank (2006) — used with permission of the authors.

Relatively few dog owners take their pets to veterinary clinics (Fielding, Mather and Isaacs 2005) and respondents interviewed at veterinary clinics can provide different answers to those interviewed in the general population (Fielding, unpublished data), so this ruled out collecting data at clinics (which was done by Carlisle-Frank and Frank). Instead, Bahamian pet caregivers who were 18 years or older were selected from a convenience sample. The same social science students were exposed to training and role play in class before conducting face-to-face interviews at numerous locations across New Providence. Locations were chosen where people typically had to wait, for example, to pay utility bills, and so the interview would not interfere with respondents' activities. A quota was imposed on the sample, to ensure that equal numbers of males and females, under 35 years and 35 years and over, were included. Thirty-five years represents the age which splits the adult population (18 years and older) into two almost equal parts (Department of Statistics 2002). Experience from previous studies (Fielding, unpublished data) has shown that without a quota, it is easy to obtain a sample which is dominated by females and under-populated with older males, compared with the national age/sex distribution of adults. As it is known that there are gender differences in human–animal interactions, it is desirable to safeguard against gender bias (Herzog 2007).

The results below refer to pet keepers, irrespective of the pets kept. (It is intended that the attitudes and actions of caregivers who have only cats or only dogs will be reported elsewhere, so that comparisons between specific classes of pet keepers can be made.) Not all questions were answered, so the sample sizes are not constant. For clarity of data presentation, we only indicate the number of respondents if the number of non-respondents was less than about 10% of the total.

Results

Of the 614 participants, there were almost equal numbers of males (309) and females (305), and those aged less than 35 years old (51%) and those aged 35 years or more (49%) (Fisher exact test, $n = 614$, $p = 0.81$). Only 10.3% of respondents lived in households without a motor vehicle.

Dogs were the most popular pet (found in 67.4% of households), followed by cats (25.9% of homes). In dog owning households, a mean of 1.92 dogs ($SD = 1.36$, $n = 405$) were kept, and in the case of cats, households kept a mean of 1.81 cats ($SD = 1.23$, $n = 157$). Fish were kept by 16.1% of respondents and birds were kept by 11.6% of respondents. Most respondents bought their pets (50.5%), while over a quarter (26.8%) received them as gifts—19.5% obtained pets from persons known to them. Adoption, strays, and free pet advertisements were the least popular external sources for acquiring pets (Table 2). The single most common reason for keeping pets was for companionship (41%). Other reasons included protection (31.1%) and to provide an interest (16.9%). The most common methods respondents used to dispose of dead pets were burying (57.5%), burning (14.7%), and putting the animal in the garbage (9.9%).

Table 2. How Bahamian and American pet caregivers obtained their pets.

	Bahamians ($n \approx 614$)	Americans ($n \approx 305$)
Purchase	50.5%	40.0%
Gift	27.0%	14.1%
Friend, family, co-worker	19.5%	23.9%
Adopt	15.3%	50.5%
Offspring of another pet	15.0%	8.5%
Stray	8.5%	35.7%
Answered an advertisement for a "free pet"	0.7%	8.2%
Other	0.2%	5.2%

American data (Carlisle-Frank and Frank, unpublished data)

Attachment to Pets

Questions related to attachment resulted in some unsolicited qualitative responses. Some respondents spontaneously stated that some questions, such as celebrating a pet's birthday, were "for white people." While most pet caregivers expressed love to their pets, this question produced mixed reactions. One respondent replied that he did not even say that he loved his wife every day let alone his pet, an animal that "could do nothing for" him. It should be noted that while 67.4% of respondents kept dogs and 25.9% kept cats, only 42.4% of respondents had had their pets neutered. In the case of dog owners, many did not conform to the legal requirement of having their animals registered—only 37.9% of respondents claimed to have registered their pets (Table 3). Caregivers who considered their pets as part of the family were more likely to allow their pets to live indoors than those who did not (64.5%, $n = 468$, and 42.5%, $n = 120$, respectively) (Fisher exact test, $n = 588$, $p < 0.001$).

Table 3. Pet-related behaviors of Bahamian and American pet caregivers.

	Bahamians ($n \approx 614$)	Americans ($n \approx 305$)
Pets permitted to live indoors with the family	59.9%	92.8%
Pets taken on family trips, walks, outings	56.9%	71.5%
Have ID on pets (where applicable)	52.0%	68.2%
Express love to pet at least once a day	51.0%	59.3%
Pets included in family photo album (if any) [1]	49.3%	86.4%
Neuter pets (where applicable)	42.4%	87.9%
Register pets (where applicable)	37.9%	86.8%
Relinquished a pet due to personal problems	34.3%	23.3%
Lost a pet in the last two years	26.4%	18.4%
Celebrate pets' birthday	22.8%	53.1%
Pets' names included with family names on greeting cards	21.6%	60.0%

[1] This number includes caregivers who had photographs of pets on cell phones and computer screen savers in our study.

American data (Carlisle-Frank and Frank, unpublished data)

Closeness to, and Disposal of, Pets

Most pet caregivers expressed satisfaction with their pets and considered them as part of the family. They identified with them, but also thought of them as property (Table 4). Feeling attached to the pet was associated with the way in which dead pets were disposed of. While burying was the most common method of disposing of dead pets, those who reported themselves as being attached or unattached to their pets were, overall, likely to choose different ways of disposing of their pets: burning was a more common choice in the attached group, while disposing of the animal in the garbage was more common in the unattached group ($\chi^2 = 23.2$, $df = 3$, $n = 607$, $p < 0.001$; Table 5).

General Attitudes towards Pets and Animals

Respondents were in general agreement that animals can feel pain and have feelings and needs. Most disapproved of chaining or caging dogs and declawing cats. Slightly more than half (56.3%, $n = 611$) thought that viewing pets as possessions was wrong, and 77.3% ($n = 613$) thought that neutering should be used to stop pet overpopulation (Table 6).

Table 4. Closeness of pet caregivers in The Bahamas and the US to their pets.

	Bahamians (n = 614)	Americans (n ≈ 305)
Satisfied (very satisfied to somewhat satisfied) with pet(s)	89.1%	91.8%
Consider pet(s) as member(s) of family	79.0%	96.4%
Feel attached to pet(s)	77.7%	94.1%
Identify with pet(s)	67.0%	90.8%
Consider pet(s) as property	62.2%	32.1%

American data (Carlisle-Frank and Frank, unpublished data)

Table 5. Level of attachment to pets and method of disposal used for dead pets.

	Level of Attachment	
Method of Disposal	Attached (n = 473)	Unattached (n = 134)
Burying	59.4%	53.7%
Burning	17.3%	6.0%
Other	15.4%	22.4%
Put in household garbage	7.8%	17.9%

Table 6. Percentage agreement of Bahamian and American caregivers with statements about pets and other animals.

	Agree with Statement	
Statement	Bahamians (n ≈ 614)	Americans (n ≈ 305)
Animals can feel pain[1]	96.9%	
Animals have feelings with needs/interests of their own[1]	95.3%	88.9%
Pets should not live long-term in cages	85.8%	89.8%
We should help animals as they are dependent on humans	85.0%	90.5%
Long-term chaining of dogs should not happen	83.0%	84.9%
Pets should be neutered to stop over-population and suffering of animals	77.3%	87.9%
Declawing cats for convenience of people in wrong	67.9%	71.1%
Viewing pets as possessions is wrong	56.3%	84.9%
We should not make a big deal out of protecting pets	21.3%	7.9%

[1]Americans were asked about agreement on a single statement: Animals are sentient beings with needs/interests of their own.

American data (Carlisle-Frank and Frank, unpublished data)

Discussion

As this study used a convenience sample, unknown biases may exist, so extrapolation to the wider population of Bahamian pet caregivers in New Providence should be made with caution. However, where points of comparison exist with other studies on animals (e.g., Fielding, Mather and Isaacs 2005; Fielding and Plumridge 2005b), broadly similar results were obtained. Also, the percentage of respondents without motor vehicles in the present study is in keeping with that found in the 2000 census (Department of Statistics 2002).

That most pets are bought probably reflects informal breeding or enterprises rather than purchases from pet shops, as only one pet shop is listed in the Yellow Pages of the 2007 telephone directory (Bahamas Telephone Company n.d.), and many advertisements for dogs appear in local newspapers (Fielding, Mather and Isaacs 2005). The breeding of animals such as cats and dogs is not subject to regulation under the present laws. A smaller percentage of Bahamians than Americans adopted pets or obtained free ones. This is of interest, as there appears to be a large number of unwanted animals available from the local humane society, through animal welfare groups, and off the street. While this unwillingness to adopt pets in a community which has a dog overpopulation problem (Fielding, Mather and Isaacs 2005) may be of concern to animal welfare groups, a similar reluctance was found in Dominica (Alie et al. 2007). Unlike elsewhere (e.g., Frank and Carlise-Frank 2003), factors which motivate adoption of shelter animals have yet to be investigated in The Bahamas. However, the source of street dogs may be more limited than has been commonly believed, as most, if not all, dogs roaming in residential areas of Nassau probably have caregivers (Fielding 2008).

Many of the pet caregivers in New Providence appreciated the welfare needs of animals, even though they were almost evenly split with regarding pets as possessions. With both laws and animal welfare groups referring to pet keepers are "owners," it is understandable that many consider pets as property.

With respect to general animal welfare issues, both Bahamians and North Americans responded similarly. They were equally averse to the declawing of cats and we hope that this general intolerance arises from interactions with veterinarians who would be aware of the potential problems associated with onychectomy in cats (Patronek 2001). However, the findings from both this study and Carlisle-Frank and Frank (2006) are at variance with the results from a Canadian study (Landsberg 1991)—possibly there has been a change in view towards onychectomy since 1991. Caregivers in both studies were generally against the chaining of dogs. It would be of interest to know if this intolerance is due to caregivers being aware of the dangers to others and to dogs of keeping dogs chained (Humane Society of the United States n.d.), or if other factors are responsible.

In both studies, cats and dogs were the most popular pets. While respondents kept a similar number of cats per caregiver in both studies (1.8 cats in the Bahamas and 1.6 in North America), caregivers in The Bahamas kept more dogs than North American caregivers (1.9 vs. 1.4 dogs per person). It would be of interest to investigate why households in The Bahamas tend to have more dogs than American households and to discover the factors which influence dog numbers within households.

While Bahamians appear to be almost as satisfied with their pets as Americans, the pet is less well integrated into the Bahamian household than the US household, and Bahamians are less likely than Americans to identify with their pets. The present study showed that caregivers who considered pets as part of the family were more likely to keep their pets in the home and so allow for greater pet–caregiver interaction. Attachment was associated with the method of disposal of pets, as has been reported elsewhere (Varner and Johnson Varner 1983). Lack of attachment could result in caregivers viewing dead pets as trash, rather than treating them respectfully in death. The lower level of interaction between Bahamian caregivers and pets further suggests that West Indians are "passive owners," compared with American caregivers. As such, they may be forgoing the beneficial effects of close interaction with pets (Beck and Katcher 1996).

Differences between Bahamian and American caregivers with regard to affection and attachment could ultimately give American visitors the impression that Bahamians do not care as much for their pets as they do. These differences, combined with the presence of roaming animals, might be responsible for the negative reactions such visitors have when they visit (Plumridge and Fielding 2003). While these reactions may be another demonstration of differing cultural expectations of tourists from their hosts (e.g., Armstrong et al. 1997), they may be ultimately detrimental to The Bahamas: some Americans may cease to visit because they perceive the treatment of animals to be poor.

Despite the acceptance that The Bahamas suffers from a dog overpopulation problem (Fielding, Mather and Isaacs 2005), Bahamians are less likely than Americans to consider that neutering is the way to control the pet population. This response is made more interesting because there are well publicized free spay/neuter programs in New Providence (e.g., The Tribune 2007). If the American model of promoting high neutering rates to reduce the pet population (Humane Society of the United States 2007) is to be adopted in The Bahamas, it is clear that there is some way to go. However, it should be noted that even within the United States, important regional variations exist with respect to neutering rates, and these may be ethnically based (Poss and Bader 2007). The high percentage of all types of Bahamian caregiver who thought that animals have "feelings with needs/interests of their own" may reflect why Bahamians are generally against getting their pets neutered—in the case of dogs, Bahamians appear to project their sexuality on to their pets (Fielding, Samuels and Mather 2002). If this is so, efforts to educate caregivers on the importance of neutering pets should be paramount.

The under-representation of black people working in animal shelters in North America has been noted as a cause for concern (Brown 2005a), and Shore et al (2006) noted some possible ethnic differences in the likelihood of a dog being kept outside (black people were more likely than white people to keep dogs in their yards). In the present study, ethnic differences arose obliquely when some respondents considered some questions as being "for white people" (e.g., concerning the celebration of a pet's birthday). (It should be remembered that in this study Bahamians interviewed Bahamians and so no white person was present during the interview, which ensured the integrity of any cultural sensitivity.) Clearly, some respondents had preconceived ideas as to how white people and black people care for pets. The results from the present study did not always support apparently commonly held beliefs regarding pet care in The Bahamas.

The college students who had been consulted when the survey was being designed had suggested that the question about celebrating a pet's birthday should be omitted, as they "knew" that no one in The Bahamas would do this. However, the present study showed that some Bahamians do celebrate their pets' birthdays and include pets in their photograph albums. Several students reacted to these results by saying that it was not "normal," which suggests deep-seated opinions concerning activities with pets. These views are noteworthy, as they were voiced by a group of people who will eventually represent the more educated members of society. The issues underlying these views need further investigation.

This study points to differences between Bahamian and North American communities with respect to pet care, but it is not possible to analyze what might be underpinning the differences. The comparison may be highlighting cultural differences or socio-economic or developmental differences based on education and income; issues which others have noted when observing superficial differences between ethnic groups (Brown 2005b). A study using respondents matched for socio-economic differences from both communities might be useful in this regard to examine the factor(s) which distinguish the two populations. While the present study allows overall comparison of attitudes and actions of caregivers in two different, yet connected, communities, it points to the need for further studies so that the importance of cultural, as opposed to socio-economic factors, on animal welfare can be assessed.

Acknowledgements

The author is grateful to Pamela Carlisle-Frank and Joshua Frank for permission to use a modified form of their questionnaire and for making available their data for comparison in this paper.

References

Alie, K., Davis, B. W., Fielding, W. J. and Maldonado, F. G. 2007. Attitudes towards dogs and other "pets" in Roseau, Dominica. *Anthrozoös* 20: 143–154.

Armstrong, R. W., Mok, C., Go, F. M. and Chan, A. 1997. The importance of cross-cultural expectations in the measurement of service quality perceptions in the hotel industry. *International Journal of Hospitality Management* 16(2): 181–190.

Bahamas Telephone Company. n.d. The Bahamas telephone directory 2007 Yellow Pages. Nassau, The Bahamas: BTC Directory Publications.

Beck, A. and Katcher, A. 1996. *Between Pets and People.* West Lafayette, IN: Purdue University Press.

Brown, S. E. 2005a. The under-representation of African Americans in animal welfare fields in the United States. *Anthrozoös* 18(2): 98–121.

Brown, S. E. 2005b. The under-representation of African American employees in animal welfare organizations in the United States. *Society & Animals* 13(2): 153–162.

Carlisle-Frank, P. and Frank, J. M. 2006. Owners, guardians and owner-guardians: Differing relationships with pets. *Anthrozoös* 19(3): 225–242.

Craton, M. and Saunders, G. 1992 and 1998. *Islanders in the Stream: A History of the Bahamian People.* Two volumes. Athens, GA: University of Georgia Press.

Department of Statistics. 2002. Report of the 2000 census of population and housing. Ministry of Economic Development, Nassau, The Bahamas.

Fielding, W. J. 2007. Knowledge of the welfare of non-human animals and prevalence of dog car practices in New Providence, The Bahamas. *Journal of Applied Animal Welfare Science* 19(2): 153–168.

Fielding, W. J. 2008. Dogs: A continuing and common neighborhood nuisance of New Providence, The Bahamas. *Society & Animals* 16(1): 61–73.

Fielding, W. J., Mather, J. and Isaacs, M. 2005. *Potcakes. Dog Ownership in New Providence, The Bahamas.* West Lafayette, IN: Purdue University Press.

Fielding, W. J. and Plumridge, S. 2004. Preliminary observations on the role of dogs in household security in New Providence. *Anthrozoös* 17(2): 167–178.

Fielding, W. J. and Plumridge, S. 2005a. Letter to the editor. *Bahamas Journal of Science* 12(2): 20.

Fielding, W. J. and Plumridge, S. 2005b. Characteristics of owned dogs on the island of New Providence, The Bahamas. *Journal of Applied Animal Welfare Science* 8(4): 245–260.

Fielding, W. J., Samuels, D. and Mather, J. 2002. Attitudes and actions of West Indian dog owners towards neutering: A gender issue? *Anthrozoös* 15(3): 206–226.

Frank, J. M. and Carlisle-Frank, P. 2003. Attitudes and perceptions regarding pet adoption. Annual meeting of the American Sociological Association.<http://convention.allacademic.com/asa2003/view_paper_info.html? pub_id=665andpart_id1=45126> Accessed May 31, 2007.

Herzog, H. A. 2007. Gender differences in human–animal interactions. *Anthrozoös* 20(1): 7–21.

Humane Society of the United States. n.d. Do you chain your dog? <http://www.hsus.org/pets/pet_care/dog_care/do_you_chain_your_dog.html> Accessed May 11, 2007.

Humane Society of the United States. 2007. Spay day USA spotlights plight of pet overpopulation and homelessness. 15th February 2007. <http://www.hsus.org/press_and_publications/press_releases/spay_day_usa_spotlights.html> Accessed March 21, 2008.

Johnson, P., Ballance, V., Fielding, W. J., McDonald, T., Scriven, C. and Stuart, M. 2005. Haitian Migration in The Bahamas. 2005. Submitted to the International Organization for Migration: Washington, DC.

Langsberg, G. M. 1991. Cat owner's attitudes towards declawing. *Anthrozoös* 4(3): 192–197.

Patronek, G. J. 2001. Assessment of claims of short- and long-term complications associated with onychectomy in cats. *Journal of the American Veterinary Medical Association* 219(7): 932–937.

Plumridge, S. and Fielding, W. J. 2003. Reactions of American tourists to roaming dogs in New Providence, The Bahamas. *Anthrozoös* 16(4): 360–366.

Poss, J. E. and Bader, J. O. 2007. Attitudes towards companion animals among Hispanic residents of a Texas border community. *Journal of Applied Animal Welfare Science* 10(3): 243–253.

Proud Paws. n.d. Proud Paws. <http://www.proudpaws.org/> Accessed May 23, 2007.

Research and Statistics Division. 2005. Exit study report. Ministry of Tourism, The Bahamas.

Scholarship and Educational Loan Division. 2003. Guaranteed loan programme. Report for the years 2000–2003. Ministry of Education, Nassau, The Bahamas.

Shore, E. R., Riley, M. L. and Douglas, D. K. 2006. Pet behaviors and attachment to yard versus house dogs. *Anthrozoös* 19(4): 325–334.

Smith, J., Singh, C., Oliver, H. and Sarjudas, S. 2006. Stray dogs. EDU 421 term paper. The College of The Bahamas.

Strachan, I. G. 2007. Going back ta da islan': Migration, memory and te marketplace in Bahamian art. *Yinna* 2: 29–46.

The Tribune. 2007. 4,000 neutered in Proud Paws project. *The Tribune*, 4 October, p.12.

UNDP. 2006. Human development report. 2006. <http://hdr.undp.org/hdr2006/statistics/> Accessed May 22, 2007.

US Census Bureau. 2001. Table FBP-1. Profile of selected demographic and social characteristics: 2000. <http://www.census.gov/population/cen2000/stp-159/stp159-bahamas.pdf> Accessed September 15, 2007.

US Department of State. 2007. Background note: The Bahamas. <http://www.state.gov/r/pa/ei/bgn/1857.htm> Accessed May 22, 2007.

Varner, J. G. and Johnson Varner, J. 1983. *Dogs of the Conquest.* Norman, OK: University of Oklahoma Press.

World Travel and Tourism Council. 2007. Tourism satellite tourism accounting tool. <http://www.wttc.travel/eng/WTTC_Research/Tourism_Satellite_Accounting_Tool/index.php> Accessed May 20, 2007.

Human-Animal Interaction: Impacting Multiple Species

"City of Fountains"

October 21-25, 2009 · Kansas City, MO USA
The Westin Crown Center   1 E. Pershing Road

The ISAZ/HAI sequential conferences will provide an exciting opportunity for those working in HAI research and practice to share their latest findings and program outcomes.

**18th Annual Conference
International Society
for Anthrozoology**

http://www.isaz.net

October 21-23, 2009

**1st Annual Conference
Research Center for
Human-Animal Interaction**

University of Missouri
College of Veterinary Medicine

http://rechai.missouri.edu

October 23-25, 2009

Research Center for Human-Animal Interaction

Principal Founding sponsors:

Founding sponsors:
Nestlé PURINA

Hosted by:

The conference brochure and registration will be posted on the web sites

The Relationship between Childhood Cruelty to Animals and Psychological Adjustment: A Malaysian Study

David Mellor*, James Yeow†, Norul Hidayah bt Mamat† and Noor Fizlee bt Mohd Hapidzal†

* School of Psychology, Deakin University, Victoria, Australia
† School of Social Science & Liberal Arts, University College Sedaya International, Kuala Lumpur, Malaysia

Address for correspondence:
Dr David Mellor,
School of Psychology,
Deakin University,
Burwood Campus, Burwood,
Victoria 3125, Australia.
E-mail: mellor@deakin.edu.au

ABSTRACT In Western research, cruelty to animals in childhood has been associated with comorbid conduct problems and with interpersonal violence in later life. However, there is little understanding of the etiology of cruelty to animals, and what in the child's life may require attention if the chain linking animal cruelty and later violence is to be broken. The study reported in this paper investigated the association between parent-reported cruelty to animals, and parent- and self-reported psychological strengths and weaknesses in a sample of 379 elementary school children in an Eastern context, Malaysia. No gender differences were found in relation to cruelty to animals or psychological problems, as assessed with the Strengths and Difficulties Questionnaire (SDQ). However, there were different predictors of cruelty to animals for boys and girls. Regression analyses found that for boys, parent-reported hyperactivity was a unique predictor of Malicious and Total Cruelty to animals. For girls, self-reported conduct problems was a unique predictor of Typical Cruelty to animals. Parent-reported total difficulties were associated with Typical, Malicious, and Total Cruelty to animals. We suggest that routine screening of children with an instrument such as the SDQ may help to detect those children who may need to undergo further assessment and perhaps intervention to break the chain linking childhood cruelty to animals and later conduct problems.

Keywords: cruelty to animals, psychological problems, Malaysia

A substantial body of Western research over the last 15 years has focused on cruelty to animals in childhood. A particular focus of this research has been on the proposed relationship between childhood cruelty to animals and subsequent violence toward other people. As early as 1961, MacDonald suggested that cruelty to animals is one of three childhood characteristics (the others being enuresis and fire setting) that may be predictive of later human violence. At about the same time, the anthropologist Margaret Mead (1964) suggested that childhood cruelty to animals is symptomatic of a violent personality that if not treated could lead to "a long career of episodic violence and murder" (p. 22). Indeed, there is now a general consensus that there is an association between childhood cruelty to animals and subsequent violence towards others, both in childhood and adulthood (Tapia 1971; Rigdon and Tapia 1977; Felthous and Kellert 1986; Tingle et al. 1986), and that by studying cruelty to animals, social scientists have "an opportunity to identify behavior that might indeed be a precursor of violence against humans to follow" (Merz-Perez, Heide and Silverman 2001, p. 571).

Empirical studies that have attempted to identify the link between childhood cruelty to animals and later violence and antisocial behaviors have been based largely on retrospective reports of violent offenders, or on case studies and anecdotal evidence. Indeed, Duncan and Miller (2002) suggested that the link between childhood cruelty to animals and violence in adulthood is based on 35 years of research using case studies of violent offenders. Many serial murderers and mass murderers were reportedly cruel to animals in their childhood, providing a strong argument for the association. However, these cases represent extreme examples of adult violence, and more systematic studies are required.

Other studies by Kellert and Felthous (1985) and Felthous and Kellert (1987) have found that violent criminals other than murderers also report a history of animal abuse. Arluke et al. (1999) compared the criminal records of animal abusers and control participants. The animal abusers were more likely to commit a range of other antisocial offences, including interpersonal violence. However, although there was an association between cruelty to animals and a variety of antisocial behaviors, the association was not limited to violence alone.

In a study of a non-criminal population, Henry (2004) asked 169 college students to self-report both observation of, and participation in, acts of cruelty to animals, as well as both recent and previous acts of delinquency. There was a strong association between these variables, with those who reported participating in or observing cruelty to animals also reporting greater involvement in a variety of delinquent behaviors both recently and over the course of their lives.

If cruelty to animals is a precursor of violence in adulthood, there is a strong argument to study it further in the childhood phase. In this endeavor, a variety of researchers (e.g., Offord, Boyle and Racine 1991; Ascione 1993; Frick et al. 1993) have noted that cruelty to animals is a symptom of childhood Conduct Disorder, and is associated with other violent and antisocial behaviors that comprise Conduct Disorder. As Henry (2004) suggests, cruelty to animals tends to occur "within the matrix of other [concurrent] antisocial behaviors" (p. 200). Consistent with this, Miller (2001) suggested that cruelty to animals may be exhibited by 25% of conduct disordered children. Similarly, Luk et al. (1999) found that children described as being cruel to animals by their parents were more likely to experience severe conduct problems. They suggested that cruelty to animals may be stable and prognostic over childhood and adolescence. Indeed, it may be the earliest symptom of conduct disorder, with Dadds, Whiting and Hawes (2006) arguing that cruelty to animals may be an early manifestation of the sub-group of children developing conduct problems associated with traits of low empathy and callous disregard for others.

Seminal works by Ascione (1992, 1993) suggested that cruelty to animals in childhood may be reflective of the development of attitudes that reflect a general callous insensitivity toward the well-being of others, both humans and animals. If such attitudes persist, the individual is prone to violence in later life. Tallichet and Hensley (2005) suggested that socialization experiences may be core to the nature of cruelty to animals. In one study, they compared 261 male inmates with rural and urban backgrounds, and reported that those from a rural background were affected by witnessing family members' abuse animals. They tended to be cruel to cats. On the other hand, those with an urban background learned to be cruel from family and friends, and were cruel to dogs, cats, and wild animals. In another study of the same inmates, Hensley and Tallichet (2005) reported that earlier observations and experiences of cruelty to animals were associated with recurrent cruelty to animals.

In a more specific study, Flynn (1999a) suggested that father's corporal punishment was related to cruelty to animals among male children, even when controlling for child abuse, father-to-mother violence, and father's education. In a similar vein, Currie (2006) reported that cruelty to animals is associated with exposure to male-to-female domestic violence. In contrast, however, Dadds, Whiting and Hawes (2006) reported that in their community sample of 151 primary school children, family problems were not associated with cruelty to animals.

In summary, as Duncan and Miller (2002) suggested, cruelty to animals in childhood "can be seen as a symptom or signal of something in the child's life that may need clinical attention" (p. 366). The current study set out to explore what in the child's life may require such clinical attention—is it simply that the child has conduct problems, or are there other areas of psychological functioning that might be of concern? To do this, we assessed various domains of childhood mental health and behavior problems, using the parent and self-report version of the Strengths and Difficulties Questionnaire, and cruelty to animals as reported by parents. Further, in contrast to most other studies which have been conducted in Western contexts, the research reported here was conducted in an Eastern context, Malaysia, where no studies on cruelty to animals have been reported. This allowed us to investigate whether childhood cruelty to animals is also associated with conduct problems, or other mental health and behavioral problems, in non-Western cultural contexts.

Methods
The Setting
Malaysian society is heterogeneous, with the Malays or the *Bumiputera* people representing 65% of the population of almost 25 million, and Chinese (26%) and Indian (8%) making up substantial minorities. By constitutional definition, all Malays are Muslim. Buddhism and Christianity, respectively, are the two next most popular religions, accounting for most of the Chinese population, although some Chinese follow Islam or Confucianism. The majority of Indian Malaysians are Hindu.

Interactions with animals are in part determined by religion. For example, while many Chinese families have dogs to guard their homes, traditionally, in Islam dogs have been seen as unhygienic, and the Islamic tradition has developed several injunctions that warn followers against most contact with dogs. While it is not forbidden to own a dog, it is not permissible to keep a dog in the house. Similarly, while it is not forbidden to touch animals, including dogs, if the saliva of a dog touches a person or any part of their clothing, then it is required that the touched body part and the item of clothing that was touched by the dog's mouth or snout be washed. It is necessary for every Muslim who owns animals, whether for farming/work purposes or as pets, to provide adequate shelter, food, water, and, when needed, veterinary care for their animals. *Haraam* (forbidden) behaviors include keeping animals in cages so small that they cannot behave naturally, keeping a dog or any other animal on a short lead for long periods without food or water, and participating in "blood sports" like dog fighting and trophy hunting (see Al-Kaysı 1986; Banderker n.d.).

Malaysia has a Society for Prevention of Cruelty to Animals, and that society has been campaigning for changes to the Animal Ordinance of 1953 to make the penalties for cruelty to animals stronger. The law at present allows a maximum penalty for cruelty to animals of RM200 (approximately USD55) and/or six months imprisonment. It does not allow for bans on ownership of animals following conviction for cruelty, and cannot prevent the return to the perpetrator of animals upon which cruelty has been inflicted. There appears to be no data on pet ownership in Malaysia, but data on pet ownership were recorded in the data collection phase of the current study. Thirty-one percent of the children were reported to currently have pets, and where the type of pet was specified (although not asked for), the majority were fish, catfish, cats, turtles and dogs.

Participants

A total of 496 Malaysian children aged between 6 and 12 years and their parents participated in a survey. The majority of the children were of Malay descent (379, 78.6%), with the remainder being Chinese (41) and Indian (62). Fourteen participants did not identify their race. In the analyses that are reported below, only the Malay participants are considered, due to the relatively small numbers of Chinese and Indian children in the sample and possible cultural differences related to animal-related attitudes and behaviors. With the removal of the Chinese and Indian children from the analyses, there were 148 Malay boys (mean age = 8.80 years, SD = 1.71 years), and 231 girls (mean age = 9.22 years, SD = 1.80 years) in the final sample.

Materials

The parent and self-report versions of the Strengths and Difficulties Questionnaire (Goodman 1997) were used to assess child mental health and behavior problems. Each version of the instrument consists of five, five-item subscales that assess conduct problems (e.g., often lies or cheats), hyperactivity-inattention (e.g., constantly fidgets and squirms), emotional symptoms (e.g., is nervously clingy in new situations), peer problems (e.g., is rather solitary, tends to play alone), and prosocial behavior (e.g., is considerate of other people's feelings). A three-point response format is used to record respondents' level of agreement that the item describes the child's behavior over the past 6 months. Scores are allocated on the basis of 0 for items checked as being "not true," 1 for items checked as being "somewhat true," and 2 for items checked as certainly true. A small number of negatively-worded items are reverse scored. Each of the five subscales is scored by adding the responses to the constituent items, and a total difficulties score is obtained by adding the scores for all but the prosocial behavior subscale. The possible range of scores for each of the subscales is 0–10, and for the total difficulties score it is 0–40. The SDQ does not include items related to cruelty to animals.

According to Goodman (2001), the items and subscales are based on current nosological concepts (e.g., the DSM-IV [American Psychiatric Association 1994] diagnostic category criteria) and factor analysis. The SDQ has been found to be comparable with, or better than, the Child Behavior Checklist (Achenbach 1991a, 1991b, 1991c) in detecting disorders when compared with a standardized semi-structured interview (Goodman and Scott 1999), and several other studies on community and clinical populations have demonstrated its validity (e.g.,

Smedje et al. 1999; Goodman, Renfrew and Mullick 2000; Klasen et al. 2000; Koskelainen, Sourander and Kaljonen 2000; Goodman et al. 2003; Muris, Meesters and van den Berg 2003). These studies have also demonstrated good internal reliability for the subscales and total difficulties scale across the informant versions, and sound inter-informant reliability. Other studies (Goodman 2001; Muris, Meesters and van den Berg 2003; Mellor 2004) have found the SDQ to have good test-retest reliability across the parent, teacher, and child versions.

The Children's Attitudes and Behaviours towards Animals questionnaire (CABTA, Guymer et al. 2001) was used to assess childhood cruelty to animals. The CABTA is completed by parents, who respond to 24 questions that are spread over three sections. Section A consists of six general demographic questions and basic pet ownership questions. Section B is made up of eight general questions which relate to the child's attitudes and behaviors towards animals (e.g., "My child has ridden a horse," "My child is afraid of animals," and "My child enjoys spending time with animals"). Section C consists of 13 questions which relate specifically to cruelty to animals. These items were generated by reference to Ascione, Thompson and Black's (1997) nine dimensions of cruelty to animals. Parents respond yes (score = 1) or no (score = 0) to four items "My child has harmed small insects," "My child has harmed other non-domestic animals," "My child has harmed other people's pets," "My child has harmed his/her own pets," and on a four-point scale anchored at never (score = 0) and always (score = 4) to 9 other items such as "My child is rough with animals," "My child has shown concern for the suffering of animals," and "My child has shown pleasure when harming animals."

The CABTA yields scores on two subscales: Typical Cruelty (possible range 0–21) and Malicious Cruelty (possible range = 0–17). Typical Cruelty involves being rough with animals, harming one's own pets, harming animals alone etc., perhaps out of curiosity. Malicious Cruelty on the other hand involves intentionally or secretly harming animals, harming animals with others, and taking pleasure in harming animals. A total score (Total Cruelty) is also obtained by summating the Typical and Malicious scores and subtracting the score for one item that loads on each subscale (possible range = 0–34). Guymer et al. (2001) have shown the CABTA to be both reliable (within subscales and overall, and test-retest) and valid against a semi-structured interview with parents regarding their child's behavior toward animals.

All questionnaires were translated into Bahasa Malaysia by a Malaysian post-graduate psychology student studying in Australia who was fluent in both English and Bahasa Malaysia. The instruments were then back-translated by one of the researchers (JY) and another bi-lingual student in Malaysia, and adjustments made to the original translations as appropriate.

Procedure
Permission to undertake the study was obtained from the Research Committee at University College Sedaya, the Educational Planning and Research Division of the Malaysian Ministry of Education, and the Jabatan Pelajaran Negri Selangor (State of Selangar Education Department). Three primary schools in the areas of Damansara, Subang Jaya, and Kuala Selangor were selected with the aim of recruiting a sample that was diverse with regard to urban and rural representation, socio-economic status, and ethnicity. Principals of those schools were approached by telephone and invited to participate in the study. All agreed to facilitate the study. The researchers then visited the schools to explain the project in more detail.

School principals distributed materials through one class at each year level. A pack that included an invitation to parents to participate in the research and to allow their child to participate was sent home with each child in the selected classes. If parents were willing to participate and have their child participate, they completed the parent version of the SDQ and the CABTA, and returned them in a sealed envelope to the school. Children of these parents were then invited to complete the child version of the SDQ in the classroom on a voluntary basis. The researchers then collated the questionnaires via the birth-date, child's gender, and classroom. The response rate was about 90%.

Analyses

Data were analyzed with the SPSS for Windows statistical package (SPSS Inc, 2003 - SPSS for Windows: Release 12.01, Michigan, IL: SPSS Inc). Before analyses were conducted, the data were checked for missing values. The few missing values (0.2%) were randomly spread. As a result, they were replaced using the mean score for that item. Skewness and kurtosis testing indicated that SDQ subscales were normally distributed. The CABTA subscales did exhibit large skewness and kurtosis values; however, these were not transformed as they are considered to be an accurate reflection of participants' responses. Scatterplots were utilized to examine the data for the assumption of linearity, which was not violated. There also did not appear to be any significant outliers. As such, the data were considered to fulfill the assumptions for correlation. Correlations and regression analyses were used to analyze the data.

Results

Table 1 provides a summary of the descriptive data relevant to the variables of interest. An Analysis of Variance indicted that there were no significant gender differences for any of the variables.

Table 1. Means, standard deviations and ranges of scores for girls and boys on each scale.

	Girls ($n = 238$)			Boys ($n = 141$)		
	Mean	SD	Range	Mean	SD	Range
Malicious Cruelty	1.28	1.41	0–9	1.46	1.58	0–9
Typical Cruelty	1.84	1.83	1–11	2.12	2.37	0–12
Total Cruelty	2.84	2.44	1–16	3.26	3.15	0–19
Parent reported emotional symptoms	2.27	2.00	0–9	1.93	1.87	0–9
Parent reported conduct problems	1.84	1.55	0–9	1.62	1.43	0–6
Parent reported hyperactivity/inattention	2.95	2.09	0–9	3.18	1.94	0–10
Parent reported peer problems	2.31	1.53	0–7	2.30	1.52	0–8
Parent reported prosocial behavior	7.84	1.87	1–10	7.74	1.81	3–10
Parent reported total difficulties	9.37	5.27	0–24	9.03	4.65	0–23
Self reported emotional symptoms	2.93	2.07	0–10	2.57	2.05	0–10
Self reported conduct problems	1.75	1.56	0–8	1.83	1.51	0–6
Self reported hyperactivity/inattention	2.69	1.92	0–9	2.77	1.94	0–10
Self reported peer problems	2.31	1.73	0–10	2.45	1.45	0–7
Self reported prosocial behavior	7.35	1.89	0–10	6.80	1.90	2–10
Self reported total difficulties	9.68	5.09	0–30	9.62	4.91	0–24

Correlations were conducted between the subscales and total scores of the SDQ and CABTA questionnaires. Results indicated that both parent- and child-reported SDQ subscales correlated variously and significantly with the CABTA scales (see Table 2). Multiple regression analyses were then conducted to ascertain which psychological variables were most predictive of the various forms of cruelty. These were conducted separately for boys and girls because different patterns of correlation were observed in the preceding analysis (Table 2). In the regressions, only those variables that were found to correlate significantly with the cruelty scales were included, and the total difficulty scores were not included because they are a composite of the four difficulties subscales. Further, analyses were conducted only if more than one psychological variable was significantly correlated with the cruelty scale. Thus for girls, no analysis was conducted for Malicious Cruelty, and for boys, no analysis was conducted for Typical Cruelty. However, parent-reported hyperactivity/inattention was significantly correlated with each of these.

Table 2. Correlations between the SDQ subscales and the CABTA scales, for girls and boys.

	Girls (n = 238)			Boys (n = 141)		
	Malicious Cruelty	Typical Cruelty	Total Cruelty	Malicious Cruelty	Typical Cruelty	Total Cruelty
Parent reported emotional symptoms	0.12	0.24**	0.18**	0.18*	0.15	0.16
Parent reported conduct problems	0.12	0.23**	0.20**	0.12	0.14	0.13
Parent reported hyperactivity/inattention	0.21**	0.26**	0.23**	0.29**	0.22*	0.25**
Parent reported peer problems	0.00	0.17*	0.11	0.07	0.07	0.08
Parent reported prosocial behavior	–0.07	–0.23**	–0.12	–0.11	0.02	–0.03
Parent reported total difficulties	0.17*	0.31**	0.25**	0.26**	0.22*	0.24**
Self reported emotional symptoms	0.07	0.09	0.07	0.18*	0.09	0.14
Self reported conduct problems	0.09	0.23**	0.16*	0.21*	0.16	0.18*
Self reported hyperactivity/inattention	0.04	0.15*	0.10	0.12	0.11	0.09
Self reported peer problems	–0.04	0.08	0.03	0.14	0.07	0.10
Self reported prosocial behavior	0.07	–0.11	–0.02	0.04	0.07	0.06
Self reported total difficulties	0.06	0.19**	0.13	0.23**	0.15	0.17*

** Correlation is significant at the 0.01 level (2-tailed). * Correlation is significant at the 0.05 level (2-tailed).

Table 3 summarizes the regression analysis for girls for Typical Cruelty. Together, the predictor variables included accounted for 13% of the variance ($R^2 = 0.13$, $p < 0.001$). However, the only significant individual predictor was self-reported conduct problems, which accounted for 2% of the variance.

Table 3. Psychological attributes predicting Typical Cruelty for girls.

	Typical Cruelty		
	R^2	β	sr^2
	0.13***		
P Emotional symptoms		0.11	0.01
P Conduct problems		0.03	0.00
P Hyperactivity		0.12	0.01
P Peer problems		0.08	0.01
P Prosocial behavior		–0.13	0.01
S Conduct problems		0.16*	0.02
S Hyperactivity		–0.05	0.00

*$p < 0.05$, ***$p < 0.001$; P = parent report, S = self report

Table 4 summarizes the results of the regression analysis for Malicious Cruelty for boys. The variables in combination accounted for a significant amount (13%) of the variance ($R^2 = 0.13, p < 0.01$), and parent-reported hyperactivity was a significant unique predictor, contributing 5% of the variance.

Table 4. Psychological attributes predicting Malicious Cruelty for boys.

	Malicious Cruelty		
	R^2	β	sr^2
	0.13**		
P Emotional symptoms		0.06	0.00
P Hyperactivity		0.25**	0.05
S Emotional symptoms		0.06	0.00
S Conduct problems		0.12	0.01

** $p < 0.01$; P = parent report, S = self report

Tables 5 and 6 show the regression analyses for Total Cruelty scores for girls and boys. For girls, the variables in combination accounted for a significant proportion (7%) of the variance in Total Cruelty scores ($R^2 = 0.07, p < 0.01$). However, there were no significant unique predictors. For boys, parent-reported hyperactivity and self-reported conduct problems in combination accounted for 9% of the variance in Total Cruelty scores ($R^2 = 0.09, p < 0.01$), but only parent-reported hyperactivity was a significant unique predictor, accounting for 6% of the variance.

Table 5. Psychological attributes predicting Total Cruelty for girls.

	Total Cruelty		
	R^2	β	sr^2
	0.07**		
P Emotional symptoms		0.06	0.00
P Conduct problems		0.06	0.00
P Hyperactivity		0.15	0.01
S Conduct problems		0.09	0.01

** $p < 0.01$; P = parent report, S = self report

Table 6. Psychological attributes predicting Total Cruelty for boys.

	Total Cruelty		
	R^2	β	sr^2
	0.09**		
P Hyperactivity		0.26**	0.06
P Conduct problems		0.13	0.02

** $p < 0.01$; P = parent report, S = self report

Discussion

This study aimed to determine whether in a Malaysian sample childhood cruelty to animals would be associated with conduct problems, as has been demonstrated in Western studies, and also with other mental health and behavioral problems. Our findings indicate that the boys and girls in our sample did not differ with regard to any of the domains of psychological functioning assessed, nor in cruelty to animals. However, there were different patterns of association between the variables of interest.

The finding of no gender difference for any of the cruelty to animals scales contrasts with findings reported elsewhere across the age span of childhood to early adulthood. For example, relying on parent reports of the behavior of clinical and control child participants in the United States, Achenbach et al. (1991) found that boys from each group were cruel to animals at a greater rate than were girls. Dadds et al. (2004) replicated this finding in community samples of Australian children. Similarly, Tapia (1971) and Rigdon and Tapia (1977) reported that in clinical samples of youth, males were more cruel to animals than females, and Miller and Knutson (1997) and Flynn (1999a, 1999b, 2000) found that animal cruelty is about four times more common in the history of male university students than female university students. It is not clear why no gender difference was found in our study. The possibilities are that boys were less cruel than in other contexts, that girls were more cruel, or both. In any of these cases, it could be that socialization, cultural and religious factors play some role, as suggested by Flynn (2001). This issue needs further investigation, and might be the focus of a cross-national study.

The correlational analysis found that various psychological characteristics were associated with cruelty to animals, but in different combinations for boys and girls. More variables were associated with Typical Cruelty than Malicious Cruelty. These mixed findings lend some support to the notion that cruelty to animals is part of a matrix of antisocial behavior, or one possible indicator of broader social/emotional problems.

To investigate which of the psychological variables contributed the most variance to animal cruelty we then conducted regression analyses. Overall, for boys, parent-reported hyperactivity was a significant, unique predictor of both Malicious Cruelty and Total Cruelty, and it was also significantly correlated with Typical Cruelty, although no regression analysis was conducted in relation to this variable, due to there being no other non-composite correlates. For girls, self-reported conduct problems was the only unique predictor, and that was in relation to Typical Cruelty. Again, however, parent-reported hyperactivity/inattention was significantly correlated with Malicious Cruelty but no regression analysis was conducted in relation to this variable, due to there being no other non-composite correlates.

Other researchers (e.g., Offord, Boyle and Racine 1991; Ascione 1993; Frick et al. 1993) have noted that cruelty to animals in Western contexts is a symptom of Conduct Disorder as described in the DSM-IV (American Psychiatric Association 1994). In this study, the conduct problems subscale scores of the SDQ (self-report and parent-report) were correlated with Typical and Total Cruelty scores for girls, although as reported above only self-report conduct problems uniquely predicted Typical Cruelty to animals. For boys, self-reported conduct problems, but not parent-reported conduct problems, was correlated with Malicious Cruelty and Total Cruelty, but was a unique predictor of neither. Thus, the association between conduct problems and cruelty to animals expected on the basis of Western research was present, but neither as uniform nor as compelling as expected.

The findings suggest perhaps that more attention should be paid to parent-reported hyperactivity/inattention, especially for boys. This variable was significantly correlated with all cruelty scores for both boys and girls, and was a unique predictor of Malicious Cruelty and Total Cruelty for boys. For girls, although it was not a unique predictor of any form of cruelty, as reported above, it was the only variable to correlate significantly with Malicious Cruelty.

It is notable that factor analytic studies have demonstrated that inattention/hyperactivity and conduct problems/aggression are separable but moderately to highly correlated problem areas

(Thompson et al. 1996), and there is a considerable body of literature indicating that ADHD and Oppositional Defiant Disorder/Conduct Disorder are highly co-morbid. For example, Barkley (1990, 1991) has suggested that between 54 to 67% of children with ADHD also meet the full DSM-IV diagnostic criteria for ODD, and that between 20 to 56% of children with ADHD and 44 to 50% of adolescents with ADHD have, or will go on to develop, CD. Thus, it is not unexpected that the two areas, hyperactivity and conduct problems, play out together when it comes to cruelty to animals, although there has been little direct research on ADHD and animal cruelty.

If these two externalizing behavioral difficulties—conduct problems and hyperactivity/inattentions—are indeed indicative of a propensity to be cruel to animals, then they could be an early marker for further psychological and behavioral problems in later life, since considerable research (e.g., Tapia 1971; Rigdon and Tapia 1977; Felthous and Kellert 1986; Tingle et al. 1986) has pointed out that children who are cruel to animals are more likely to generalize this cruelty to violence against other people.

Perhaps more informative was the strong association (correlation) between parent-reported total difficulties scores and each of three kinds of cruelty to animals scores for both boys and girls. This finding also provides some general support for Henry's (2004) suggestion that cruelty to animals is part of a matrix of antisocial behaviors, and the general psychological deviance hypothesis. As pointed out in the Results section, the total SDQ difficulties scores were not included in the regression analyses because they are composite scores derived from four of the five SDQ subscales and therefore would have confounded the analyses because they would co-vary with the subscale scores entered into the analyses. Despite this, it is noteworthy that cruelty to animals seems to be reliably associated with overall parental evaluations of their child's difficulties. As such, this parental evaluation could be a good predictor of a range of difficulties including cruelty to animals, which, as mentioned above, may have predictive power in relation to future anti-social/violent behaviors. This suggests that screening children periodically by means of the SDQ could be a useful strategy in the attempt to prevent difficulties in adolescence and later life, provided that opportunities for further assessment and possible intervention are available in the event that the screening is indicative of potential problems.

While these findings provide some possible indicators of cruelty to animals in children, there are several limitations to this study. Firstly, they may be culturally specific and not generalizable to other contexts or even to other groups or areas in Malaysia. Secondly, the measure of cruelty to animals is based on parent report, and it may be that parents are not aware of the degree to which some children engage in such behaviors. However, concerns have been expressed about asking children directly about cruelty to animals, and other measurement approaches also rely on parent report. Thirdly, the amount of variance explained in the regression analyses, while significant, was not high. Thus, cruelty to animals is only in small part explained by the mental health and behavioral problems included in the analyses. Since our sample was based on a community sample, it is possible that we tapped a limited range of dysfunction, which reduced the likelihood of finding stronger associations between variables. Finally, the associations between the mental health/behavioral problems and cruelty to animals are simply correlational—a longitudinal study is needed to provide a greater understanding of them.

In summary, conduct problems did not seem to be as reliably associated with cruelty to animals in our Malaysian sample as in Western studies. Rather, parent-reported hyperactivity/inattention was a more important predictor. While our preliminary study has limitations, it provides the first insights into these phenomena in the Malaysian context, and suggests that further studies may enhance our understanding.

References

Achenbach, T. M. 1991a. *Manual for the Child Behavior Checklist 4–18 and 1991 Profile*. Burlington, VT: University of Vermont Department of Psychiatry.

Achenbach, T. M. 1991b. *Manual for the Teacher's Report Form and 1991 Profile*. Burlington, VT: University of Vermont Department of Psychiatry.

Achenbach, T. M. 1991c. *Manual for the Youth Self-Report.* Burlington, VT: University of Vermont Department of Psychiatry.

Achenbach, T. M., Howell, C. T., Quay, H. C. and Conners, C. K. 1991. National survey of problems and competencies among four to sixteen-year-olds. *Monographs of the Society for Research in Child Development* 56: Serial No. 255.

Al-Kaysı, M. I. 1986. *Morals and Manners in Islam: A Guide to Islamic Ādāb.* Leicester: Islamic Foundation.

American Psychiatric Association. 1994. *Diagnostic and Statistical Manual of Mental Disorders*. 4th edn. Washington, DC: Author.

Arluke, A., Levin, J., Luke, C. and Ascione, F. 1999. The relationship of animal abuse to violence and other forms of antisocial behavior. *Journal of Interpersonal Violence* 14: 963–975.

Ascione, F. R. 1992. Enhancing children's attitudes about humane treatment of animals: Generalization to human-directed empathy. *Anthrozoös* 5: 176–191.

Ascione, F. R. 1993. Children who are cruel to animals: A review of the research and implication for developmental psychopathology. *Anthrozoös* 6: 226–247.

Ascione, F. R., Thompson, T. M. and Black, T. 1997. Childhood cruelty to animals: Assessing cruelty dimensions and motivations. *Anthrozoös* 10: 170–179.

Banderker, A. M. n.d. Dogs in Islam. <http://www.islamicconcern.com/dogs.asp> Accessed February 19, 2008.

Barkley, R. A. 1990. *Attention Deficit Disorder: A Handbook for Diagnosis and Treatment.* New York: The Guilford Press.

Barkley, R. A. 1991. Diagnosis and assessment of attention deficit hyperactivity disorder. *Comprehensive Mental Health Care* 1: 27–42.

Currie, C. L. 2006. Animal cruelty by children exposed to domestic violence. *Child Abuse and Neglect* 30: 425–435.

Dadds, M.,R., Whiting, C., Bunn, P., Fraser, J. and Charlson, J. 2004. Measurement of cruelty in children: The Cruelty to Animals Inventory. *Journal of Abnormal Child Psychology* 32: 321–334.

Dadds, M. R., Whiting, C. and Hawes, D. J. 2006. Associations among cruelty to animals, family conflict, and psychopathic traits in childhood. *Journal of Interpersonal Violence* 21: 411–429.

Duncan, A. and Miller, C. 2002. The impact of an abusive family context on childhood animal cruelty and adult violence. *Aggression and Violent Behavior* 7: 365–383.

Felthous, A. R. and Kellert, S. R. 1986. Violence against animals and people: Is aggression against living creatures generalised? *Bulletin of the American Academy of Psychiatry and Law* 14: 55–69.

Felthous, A. R. and Kellert, S. R. 1987. Childhood cruelty to animals and later aggression against people: a review. *American Journal of Psychiatry* 144: 710–717.

Flynn, C. P. 1999a. Animal abuse in childhood and later support for interpersonal violence in families. *Society & Animals* 7(2): 161–171.

Flynn, C. P. 1999b. Exploring the link between corporal punishment and children's cruelty to animals. *Journal of Marriage and the Family* 6: 971–981.

Flynn, C. P. 2000. Why family professionals can no longer ignore violence toward animals. *Family Relations* 49: 87–95.

Flynn, C. P. 2001. Acknowledging the "Zoological Connection": A sociological analysis of animal cruelty. *Society & Animals* 9: 71–87.

Frick, P. J., Lahey, B. B., Loeber, R., Tannenbaum, L., van Horn, Y., Lahey, B. B., Christ, M. A .G., Hart, E. A. and Hanson, K. 1993. Oppositional defiant disorder and conduct disorder: A meta-analytic review of factor analyses and cross-validation in a clinical sample. *Clinical Psychology Review* 13: 319–340.

Goodman, R. 1997. The Strengths and Difficulties Questionnaire: A research note. *Journal of Child Psychology and Psychiatry* 38: 581–586.

Goodman, R. 2001. Psychometric properties of the Strengths and Difficulties Questionnaire. *Journal of the American Academy of Child and Adolescent Psychiatry* 40: 1337–1345.

Goodman, R., Ford, T., Simmons, H., Gatward, R. and Meltzer, H. 2003. Using the Strengths and Difficulties Questionnaire (SDQ) to screen for child psychiatric disorders in a community sample. *International Review of Psychiatry* 15: 166–172.

Goodman, R., Renfrew, D. and Mullick, M. 2000. Predicting type of psychiatric disorder from Strengths and Difficulties Questionnaire (SDQ) scores in child mental health clinics in London and Dhaka. *European Child and Adolescent Psychiatry* 9: 129–134.

Goodman, R. and Scott, S. 1999. Comparing the Strength and Difficulties Questionnaire and the Child Behavior Checklist: Is small beautiful? *Journal of Abnormal Child Psychology* 27: 17–24.

Guymer, E., Mellor, D., Luk, E. and Pearse, V. 2001. The development of a screening questionnaire for childhood cruelty to animals. *Journal of Child Psychology and Psychiatry and Allied Disciplines* 42: 1057–1063.

Henry, B. 2004. The relationship between animal cruelty, delinquency, and attitudes toward the treatment of animals. *Society & Animals* 12: 185–207.

Hensley, C. and Tallichet, S. E. 2005. Learning to be cruel?: Exploring the onset and frequency of animal cruelty. *International Journal of Offender Therapy and Comparative Criminology* 49: 37–47.

Kellert, S. R. and Felthous, A. R. 1985. Childhood cruelty toward animals among criminals and noncriminals. *Human Relations* 38: 1113–1129.

Klasen, H., Woerner, W., Wolke, D., Meyer, R., Overmeyer, S., Kaschnitz, W., Rothenberger, A. and Goodman, R. 2000. Comparing the German versions of the Strengths and Difficulties Questionnaire (SDQ-Deu) and the Child Behavior Checklist. *European Journal of Child and Adolescent Psychiatry* 9: 271–276.

Koskelainen, M., Sourander, A. and Kaljonen, A. 2000. The Strengths and Difficulties Questionnaire among Finnish school-aged children and adolescents. *European Child and Adolescent Psychiatry* 9: 277–284.

Luk, E. S. L., Staiger, P. K., Wong, L. and Mathai, J. 1999. Children who are cruel to animals—A revisit. *Australian and New Zealand Journal of Psychiatry* 33: 29–36.

MacDonald, J. 1961. *The Murderer and His Victim.* Springfield, IL: Charles C. Thomas.

Mead, M. 1964. Cultural factors in the cause and prevention of pathological suicide. *Bulletin of the Menninger Clinic* 28: 11–22.

Mellor, D. 2004. Furthering the use of the SDQ: Reliability with younger child respondents *Psychological Assessment* 16: 396–401.

Merz-Perez, L., Heide, K. M. and Silverman, I. J. 2001. Childhood cruelty to animals and subsequent violence against humans. *International Journal of Offender Therapy and Comparative Criminology* 45: 556–573.

Miller, C. 2001. Childhood cruelty to animals and interpersonal violence. *Clinical Psychology Review* 21: 735–749.

Miller, K. S. and Knutson, J. F. 1997. Reports of severe physical punishment and exposure to animal cruelty by inmates convicted of felonies and by university students. *Child Abuse and Neglect* 21: 59–82.

Muris, P., Meesters, C. and van den Berg, F. 2003. The Strengths and Difficulties questionnaire (SDQ): Further evidence for its reliability and validity in a community sample of Dutch children and adolescents. *European Child and Adolescent Psychiatry* 12: 1–8.

Offord, D. R., Boyle, M. H. and Racine, Y. A. 1991. The epidemiology of antisocial behavior in childhood and adolescence. In *The Development and Treatment of Childhood Aggression,* 31–54, ed. D. J. Peppler and K. H. Rubin. NJ: Lawrence Erlbaum Associates.

Rigdon, J. D. and Tapia, F. 1977. Children who are cruel to animals: a follow-up study. *Journal of Operational Psychiatry* 8: 27–36.

Smedje, H., Broman, J-E., Hetta, J. and von Knorring, A-L. 1999. Psychometric properties of a Swedish version of the "Strengths and Difficulties Questionnaire." *European Child and Adolescent Psychiatry* 8: 63–70.

Tallichet, S. E. and Hensley, C. 2005. Rural and urban differences in the commission of animal cruelty. *International Journal of Offender Therapy and Comparative Criminology* 49: 711–726.

Tapia, F. 1971. Children who are cruel to animals. *Child Psychiatry and Human Development* 22: 70–77.

Thompson, L. L., Riggs, P. D., Mikulich, S. K. and Crowley, T. J. 1996. Contribution of ADHD symptoms to substance problems and delinquency in Conduct-Disordered adolescents. *Journal of Abnormal Child Psychology* 24: 325–347.

Tingle, D., Barnard, G. W., Robbins, L., Newman, G. and Hutchinson, D. 1986. Childhood and adolescent characteristics of pedophiles and rapists. *International Journal of Law and Psychiatry* 9: 103–116.

Moral and Fearful Affiliations with the Animal World: Children's Conceptions of Bats

Peter H. Kahn, Jr.*, Carol D. Saunders†, Rachel L. Severson*, Olin E. Myers, Jr. ‡ and Brian T. Gill§

*Department of Psychology, University of Washington, Seattle, USA
† Department of Communications Research and Conservation Psychology, Chicago Zoological Society, Brookfield, USA
‡ Huxley College of the Environment, Western Washington University, Bellingham, USA
§ Department of Mathematics, Seattle Pacific University, Seattle, USA

Address for correspondence:
Dr P. H. Kahn, Jr.,
Department of Psychology,
University of Washington,
Box 351525, Seattle, WA
98195-1525,
USA.
E-mail:
pkahn@u.washington.edu

ABSTRACT The purpose of this study was to extend knowledge on how children understand their affiliation with an animal that can evoke both fear and care: bats. We interviewed 120 children, evenly divided between four age groups (6–7, 9–10, 12–13, and 15–16 years) after each child had visited an exhibit at Brookfield Zoo that displays Rodrigues fruit bats. Results showed that in the same children a fear orientation toward bats existed alongside of a caring orientation. Children accorded bats the right to live free and to be wild. Yet most of the same children also said that zoos did not violate the rights of bats by keeping them in captivity. Discussion focuses on this seeming contradiction, and the resulting implications for the ecological mission of many zoos today.

Keywords: bats, biophilia, care, fear, moral development

 There is little disagreement that people affiliate positively with many aspects of nature. People enjoy walking along the ocean's edge, for example, or watching the sunrise, listening to song birds, and cuddling a dog. People also fear aspects of nature. It can be the fear of walking in woods at dark, or of encountering a rattlesnake, or of being battered by a winter storm. But is it possible that people can affiliate positively with—and bring caring and moral relationships to—an aspect of nature that they simultaneously fear?

Toward investigating this question from a developmental perspective, we interviewed children about a type of animal—bats—that we believed

would readily evoke in children both moral and fearful affiliations. Rather than ask children questions in a hypothetical context (e.g., "imagine if you saw a bat…"), we recruited our participants from children that had just exited from an exhibit at Brookfield Zoo (Brookfield, IL, USA) that displays Rodrigues fruit bats. In this mostly darkened exhibit, there was no barrier (such as a glass or wire mesh) between the bats and visitors; and, indeed, sometimes the bats swooped around the visitors, occasionally within inches. Thus, by providing children with direct, unmediated contact with live bats, we were in a strong position to pursue three central lines of investigation: caring for bats, fear of bats, and the potential moral basis for keeping or not keeping bats in captivity.

Caring for Bats

At an early age, children come to care for a wide range of animals. In a year-long study, for example, Myers (2007) documented the caring relationships that children formed with a dog, turtles, a guinea pig, goldfish, doves, ferrets, pythons, a spider monkey, bugs, and squirrels. Covert et al. (1985) found that 75% of the children in their study between ages 10 and 14 said that they turned to their pets when they were upset. Melson (2001) writes that a pet's "animate, responsive proximity makes children feel less alone in a way that toys and games, television or video, even interactive media, cannot" (p. 59). In other literature, Kahn and colleagues conducted interviews in diverse locations, including Houston, USA (Kahn and Friedman 1995), the Brazilian Amazon (Howe, Kahn and Friedman 1996), and Lisbon, Portugal (Kahn and Lourenço 2002) with children about their environmental views and values. Results showed that animals, plants, and parks and open spaces played an important role in children's lives. Children were aware that water pollution can harm birds, water, insects, and landscape aesthetics. Moreover, it mattered to children that harm might occur to each of these environmental constituents. Based on measures that controlled for magnitude of environmental harm and proximity to harm, children also believed that polluting a waterway violates a moral obligation. Thus, one important finding from the research literature is that children in diverse cultures—and even in harsh urban landscapes—have in at least certain respects meaningful and moral relations with nature.

In the current study, we investigated how children conceive of the idea of caring for a type of animal, a bat, that is neither the large, charismatic megavertebrate nor warm and cuddly, but a somewhat scary-looking animal that people can fear (Lawrence 1993). We asked children questions about whether they cared about bats and why, and what they meant by the idea of care. We expected that in various ways children would express care for bats. We also sought to characterize the different ways that children conceptualize the idea of caring. Here we expected to build on the distinction between anthropocentric and biocentric reasoning (Nevers, Gebhard and Billmann-Mahecha 1997; Kahn 1999; Kahn 2006). Anthropocentric reasoning focuses on how effects to the environment affects human beings. In other words, environmental constituents may be given consideration (e.g., "it's wrong to kill fish by polluting the waters") but the reason is human-oriented (e.g., "because I like to go fishing"). Biocentric reasoning appeals to the moral standing of nature at least partly independent of its value to humans (e.g., "animals have rights to their freedom").

Fear of Bats

The research literature has established what perhaps we know intuitively: that children can fear various aspects of nature. Bowd (1983) found that kindergarten children feared such animals as bears, tigers, snakes, lions, horses, and elephants, and children's reasons most often focused on being bitten or hurt. According to Maurer (1965, p. 265), "[a]lmost all 5- and 6-year-olds and more than one-half of 7-to-12-year-olds claim that the things to be afraid of are mammals and reptiles (most frequently): snakes, lions, and tigers." Through counter-conditioning experiments, Ulrich reports that the processing of biologically prepared fear-relevant natural stimuli (including spiders and snakes) "can be very fast and may often occur automatically or 'unconsciously'" (Ulrich 1993, p. 85). Ascione (1993) has shown that some children can be cruel to animals. Kaltenborn and colleagues (2006) report that the majority of residents around the Serengeti National Park in Tanzania favored killing lions, cheetahs, leopards, and hyenas if the animals threatened domestic livestock or people. Fears of nature extend beyond the animal form. For example, Bixler and Floyd (1997) suggest that people most often express fear of objects and situations as to their reason for negatively reacting to wild lands. Overall, Kellert (1996) largely captures this orientation under what he calls the negativistic value: an aversion, fear, and dislike of nature.

In the current study, even granting that there is a great deal of safety in experiencing animals in a zoo context, it seemed likely that some children would directly express fears about bats. We sought to assess the extent of such fears, and to characterize their quality. For example, is the fear more or less like fear they feel in a dangerous part of a city? Or is it a type of fear that they like, that makes them more alert? We expected that expressed fears would diminish with age. We also expected that we would find evidence for the coexistence of fear and caring even in young children.

Moral Basis for Keeping or Not Keeping Bats in Captivity

Treating an other as a full moral entity usually depends not only on caring about or for that entity, but on attributing to that entity certain psychological qualities, such as sentiency, agency, and free will (Regan 1983). Thus, we asked questions about whether bats had these qualities, and expected that to some degree children would affirm them. Then we assessed whether children also accorded bats rights, such as the right to live free and to be wild. If they did, then we investigated how children understood the legitimacy of zoos for keeping bats in captivity. It was on open question how children would coordinate such competing claims.

Methods

Participants

Participants were 120 children, evenly divided across four age groups: 6–7 years (13 male, 17 female), 9–10 years (15 male, 15 female), 12–13 years (15 male, 15 female), and 15–16 years (17 male, 13 female). Seventy percent of participants volunteered information about their race. Of these, 86% were Caucasian, 7% were Hispanic/Latino, 2% were African American, 1% were Asian/Pacific Islander, and 4% were Other. All participants were non-members of Brookfield Zoo. Only non-members were included in the study, since they are more representative of the general population than zoo members.

Procedures and Measures

One of the exhibits at Brookfield Zoo is the "Australia House": a darkened, cave-like enclosure, about 80 feet long, that people enter and walk through. The exhibit displays Rodrigues fruit bats. A notable feature of this exhibit is that there is no barrier between the exhibit animals and the public. Thus, as people walk through the exhibit, they not only look at and hear the bats, but experience their immediate proximity.

At the entrance to the exhibit, a researcher visually identified potential participants who appeared to meet two eligibility criteria for the study. These criteria were being within the desired age range (6–16 years) and not being part of an organized party, such as a school group or camp group. Based on these criteria, a potential participant was then chosen visually at the entrance to the exhibit, observed visually through the exhibit, and then approached after exiting the exhibit. At that junction, if the child said that he or she was a non-zoo member, was within our desired age, and was interested in participating in our study, then the researcher escorted him/her to a public zoo building for an interview. Written consent was obtained at that time from the child's parent or guardian. Researchers initially identified potential participants randomly; insofar as the younger age categories filled more quickly, researchers shifted toward targeting the older age groups.

In terms of response rate, we approached a total of 348 people. Of those, 84 did not meet the eligibility criteria (61 were zoo members and 23 fell outside of our desired age range). That left a total of 264 people. From this total, our response rate was 56% (27 pilot participants, 120 regular participants). The reasons people said that they did not want to participate in this study included that it took too much time (56 individuals), they were not interested (47 individuals), and other reasons (14 individuals).

The interview was based on structural-developmental methods. These methods were pioneered by Piaget (1929/1960), and have been successfully extended by many social-cognitive researchers to date to investigate children's conceptions of social and moral life (e.g., Turiel 1998; Helwig 2006) and children's conceptions of environmental moral issues (e.g., Kahn and Friedman 1995; Howe, Kahn and Friedman, 1996; Nevers, Gebhard and Billmann-Mahecha 1997). For a chapter-length overview of this methodology, see Kahn (1999, Chap. 5). In brief, we first generated theory-guided interview questions. As a case in point, we know from the moral-developmental literature that moral relationships include substantive considerations of care, fairness, and rights, and that underlying such considerations lie judgments that the moral entities are experiencing subjects, with thoughts and feelings (Turiel 1998). Thus, we asked specific questions about all of these constructs as they apply to bats. Each participant was asked each initial question, and then the interviewer had the freedom to follow the questions in different directions, so as to tap each participant's individual understandings.

A full list of our questions can be found in Table 1. Questions focused on the study's three major areas of investigation: Caring for bats (e.g., "Do you care about bats? Why? What does it mean to care?"); fear of bats (e.g., "How did you feel with the bats flying around?" "How do you think it would feel to sleep in a place where bats were able to fly around?"); and the moral basis for keeping or not keeping bats in captivity (e.g., "Do bats have the right to be wild? Why? Does the zoo violate the rights of bats by keeping them in the Australia house? Why?").

Coding and Reliability

The interviews were tape recorded and transcribed for analysis. We then developed a coding manual, by which we mean a systematic document that explicates how to interpret and characterize (and thereby "code") the qualitative data. The generation of this coding manual followed well established methods in developmental psychology (Kohlberg 1984; Kahn 1999). In brief, we established initial conceptual categories, based on previous psychological coding systems (e.g., Kahn and Friedman 1995; Kahn and Lourenço 2002) and philosophical theory (e.g., Rawls 1971; Rolston 1989; Scheffler 1992). We then used these categories as a rough framework to interpret the qualitative data. The data, in turn, drove substantial modifications and further conceptualizations in our system, which was then reapplied to more data in an iterative manner. This dialectical process, where theory is grounded in data, and vice-versa,

Table 1. The questions used in the structured interview.

	Eval. n^1	Just. n
Caring About Bats		
1. Do you care about bats? Why?	110	89
2. What does it mean to care?	–[2]	79
3. Do you like bats?	111	–
4. Could you love a bat as a pet?	110	–
5. Would you care if there were no bats in the world? Why?	106	64
6. Does it matter to you that you were able to see bats at the zoo?	105	–
7. Let's say you lived your whole life without ever seeing a real live bat – not in a zoo, not in the wild. Just pictures. Would that matter to you? Why?	108	–
Fear of Bats		
8. How did you feel with the bats flying around?	113	–
9. Did you think that the bats would hurt you?	113	–
10. Would you like it better if the bats were separated from you (e.g., with a wire mesh between you and the bats)? Why?	110	–
11. Some people say that they pay more attention and become more alert when they walk through this bat exhibit. What do you think? Did that happen to you?	108	–
12. How do you think it would feel to sleep in a place where bats were able to fly around?	101	–
13. Would you prefer or not prefer to sleep in a place where bats were able to fly around?	103	–
14. One person I talked with said that the fear she (he for male participant) feels with bats is similar to the fear she feels when walking down a dark city street at night. Do you feel this way? Why?	64	–
15. Another person I talked with said that the fear she (he for male participant) feels with bats is very different from the fear she feels in the city. Rather, she said she kind of likes the fear she feel with bats. Do you feel this way? Why or why not?	71	–
16. Do you think it would be harmful to you to pet a bat in the zoo?	97	–
17. Do you think it would be harmful to you to pet a bat in the wild?	110	–
The Moral Basis for Keeping or not Keeping Bats in Captivity		
18. Do bats in the Australia House have feelings? If not, why not? If so, what do you think they feel?	115	–
19. Do bats in the Australia House have thoughts? If not, why not? If so, what do you think they think?	114	–
20. If a bat got mad at another bat (for seemingly no reason at all) and bit it, do you think the bat should be blamed for doing something wrong? Why?	105	–
21. If a bat got mad at you (for seemingly no reason at all) and bit you, do you think the bat should be blamed for doing something wrong? Why?	106	–
22. If a person got mad at a bat (for seemingly no reason at all) and hit it with a stone, do you think that person should be blamed for doing something wrong? Why?	113	–
23. Do you think it is all right or not all right to keep bats in a zoo? Why?	116	–
24. Do you think bats have rights? Why? Which ones?	101	–
25. Do bats have the right not to be killed by humans? Why?	89	61
26. Do bats have the right to be wild? Why?	101	71
27. Do bats have the right to live free? Why?	106	66
28. Does the zoo violate the rights of bats by keeping them in the Australia House? Why? (If "no" to this question, and "yes" to the previous question: "Just a minute ago you said that bats have the right to live free, and now you say that it's all right for zoos to violate that right. Can you help me understand what you are thinking about?")	97	–
29. Would it be better for the bats if zoos didn't keep them at all?	93	–
30. Do you think that keeping bats in the Australia House protects these bats from becoming extinct?	96	–

[1] "n" refers to the number of participants who provided responses for each of the questions that pertained to evaluations (Eval.) and justifications (Just.). [2] A dash indicates that the question was not asked.

continued until we could satisfactorily code the initial pilot data and a set of 20 random interviews from our data set. At that point, we considered the coding manual completed, and we no longer modified it. The strength of this latter approach—where the coding manual is not derived from the entire data set—is that some initial distance is traveled within a single study toward internal replicability (cf. Thompson 1996; Kahn 1999). Three types of responses were coded: (1) evaluative responses (e.g., all right or not all right, yes or no), (2) justifications for the evaluations (e.g., an appeal to the intrinsic value of nature), and (3) content responses (e.g., "scared"). Multiple justifications were coded. Summary descriptions of the justification coding systems are presented in the results section.

A coder trained in the use of the coding manual coded 60% of the data (from the 100 interviews not used in coding manual development). A second coder trained in the coding system coded the remaining 40% of the data. A third reliability coder trained in the coding system re-coded 40% of the interviews (24 randomly selected from the 60 coded by the first coder and 16 randomly selected from the 40 coded by the second coder). Intercoder reliability was assessed using Cohen's kappa. For the evaluation questions, k = 0.78, and for the content questions, k = 0.81. For justifications at the level reported in Table 3 and in the text, k = 0.63. Two commonly referenced benchmarks that can assist one in interpreting the values of Cohen's kappa are Fleiss, Levin and Paik (2003) who rate any value of kappa over 0.75 as excellent agreement, between 0.40 and 0.75 as intermediate to good, and below 0.40 as poor, and Landis and Koch (1977) who rate a kappa of 0.81 to 1.00 as "almost perfect" and between 0.61 and 0.80 as "substantial" agreement.

Results

Nonparametric tests were used to test statistical significance of the categorical data at the $\alpha = 0.05$ significance level. In particular, Kendall's tau-b, a nonparametric correlation coefficient, was used to test for developmental trends, and the Fisher exact test was used to test for gender differences. Virtually no gender differences were found, thus the data were collapsed across gender. Age differences were found where reported. Actual sample sizes for each question are reported in Table 1.

Care for Bats

By and large, children brought forward a caring orientation toward bats. For example, 78% of the children said that they cared about bats. In addition, 80% said they liked bats, 64% said that they could love a bat as a pet, 73% said that they would care if there were no bats in the world, 84% said it mattered to them that they were able to see bats at the zoo, and 69% said it would matter to them if they lived their whole life without ever seeing a real live bat (in the zoo or in the wild).

We asked children to explain their reasons for two of the above questions: for why they cared about bats and why they would care if there were no bats in the world. From the resulting justification data, we developed a typology of care for bats, which is summarized in Table 2. We then coded all of the children's reasons based on this typology. Table 3 reports the percentage of participants that used each category. As shown in Table 3, across both questions the majority of children's justifications were anthropocentric (52% and 74%), including a focus on personal interests, human welfare, and aesthetics. To a lesser extent, children drew on justifications that were biocentric (30% and 9%), including a focus on the intrinsic value of nature and justice for nature. No developmental differences were found.

Table 2. Summary of justification coding categories for caring about bats.

Category	Definition (with Example)
1. Welfare of Nature	An appeal to effects on nature, including animals, vegetation, non-living parts of nature, species, and natural processes, without specifying whether those effects led to anthropocentric or biocentric considerations (e.g., "[I care about bats] because I think they're important; they keep the insect population down like mosquitoes").
2. Anthropocentric	An appeal based on considerations of effects on humans.
2.1. Personal	An appeal based on personal predilections, interests, and projects of self and others (e.g., "[I care about bats] because I just like animals and I like nature").
2.2. Welfare	An appeal based on an others' physical, material, psychological, and educational welfare (e.g., "instead of [people] being bit by spiders and mosquitoes and stuff, the bats will eat them").
2.3. Aesthetic	An appeal based on the preservation of the environment for the viewing or, more broadly, sensorial pleasure of humans (e.g., "[I care about bats because] they're beautiful animals and I just like them around").
3. Biocentric	An appeal to the moral standing of nature at least partly independent of its value to humans.
3.1. Intrinsic Value of Nature	An appeal that nature has value, including a focus on biological life (e.g., "because it's a living thing"), natural processes (e.g., "because they're part of the earth"), and telos of nature (e.g., "because every creature has a certain role in life").
3.2. Justice	An appeal that nature has rights, deserves respect or fair treatment, or merits freedom (e.g., "they should be out in the wild and actually have a lot of freedom").

Note: To be clear, the overarching categories are Welfare of Nature, Anthropocentric, and Biocentric. Subcategories are then presented for both Anthropocentric and Biocentric (with reliability of coding established at that level).

Table 3. Percentage of anthropocentric and biocentric justifications in response to questions about caring for bats and the rights of bats.

	Why Do/Would You Care…		Why Do Bats Have the Right…		
Justification Category	About Bats? (n = 74)	If There Were No Bats in the World? (n = 55)	Not to Be Killed? (n = 61)	To Be Wild? (n = 69)	To Live Free? (n = 64)
1. Welfare of Nature	12	18	20	17	22
2. Anthropocentric	52	74	16	0	0
2.1. Personal	26	44	3	0	0
2.2. Welfare	18	24	11	0	0
2.3. Aesthetic	8	6	2	0	0
3. Biocentric	30	9	64	83	78
3.1. Intrinsic Value	23	7	26	59	45
3.2. Justice	7	2	38	23	33

Note: Percentages may not sum to 100, due to rounding.

We also asked children more generally "What does it mean to care?" From the resulting justification data, we developed a typology comprising five overarching categories: (1) *Personal Predilections*, a conception based on the personal likes and dislikes of the child (e.g., "[What caring means is] to like them"); (2) *Physical Assistance*, a conception based on beneficial action (or the negation of harmful action) to an individual or species (e.g., "What does it mean [to care]? I deposit money and stuff for the animals; it means you try to help it out");

(3) *Psychological*, a conception based on according psychological attributes, interest, vulnerability, and emotional attachment ("[Caring means] when you're emotionally attached to a pet or something"); (4) *Generalized*, a conception based on the integration of personalized caring with other moral constructs (such as justice) that leads to principles that govern interpersonal relationships ("[Caring means] to take responsibility over that thing"); and (5) *Ecological*, a conception based on the integration of generalized caring within an ecological system of which humans may be a part ("[Caring means] to not want them to be extinct and want them to be wild, but also be able to see them, too").

Quantitative results showed the following pattern of use: Personal Predilections (10%), Physical Assistance (48%), Psychological (33%), Generalized (5%), and Ecological (3%). No developmental differences were found.

Fear of Bats

Some of the children expressed fear of bats. For example, 83% of the children said that they would prefer not to sleep in a place where bats were able to fly around; and, in such a context, 57% of the children said that they would feel fearful or slightly fearful. When asked how they felt with bats flying around them in the exhibit, 32% of the children said they were scared or nervous. Twelve percent thought the bats would hurt them in the exhibit. When asked if it would be harmful to pet a bat, 29% of the children thought it would be harmful in a zoo context, while 92% thought it would be harmful in the wild, with 19% of children specifically citing fear of attack or fear of disease in a zoo context and 42% citing fear of attack or disease in the wild. Also, 27% of the children said they would have liked it better if the bats had been separated from them in the exhibit with a wire mesh.

We then investigated the quality of that fear by comparing it to the fear children might feel when walking down a dark city street at night (Question 14). A total of 27 children said that they felt scared or nervous with the bats flying around the exhibit in response to Question 8 and also provided a valid response to Question 14. Of these children, 15 (56%) said that the two types of fear felt similar. We next provided children with an alternative account of the fear (Question 15): "Another person I talked with said that the fear she feels with bats is very different form the fear in the city. Rather, she said she kind of likes the fear she feels with bats. Do you feel this way?" A total of 29 children said that they were scared or nervous with the bats flying around and also provided a valid response to Question 15. Of these children, 13 (45%) said that they "kind of liked" the fear they felt with bats.

A developmental trend was found. As shown in Table 4, compared with the younger children, the older children were less fearful of bats, welcomed greater contact with bats, and said that they felt more alert while in the presence of bats.

Table 4. Percentage of children by age who experienced aspects of fear with bats.

Judgments about Bats in the Exhibit	6–7 Years	9–10 Years	12–13 Years	15–16 Years	*p* value (Kendall tau-b)
1. Felt scared of bats	37	37	24	10	0.006
2. Thought bats would hurt them	22	14	7	3	0.021
3. Would like physical separation from bats	35	37	28	11	0.031
4. Felt more alert in the presence of bats	54	69	73	86	0.008

We then examined the potential coexistence of fear and caring. To assess this, we looked at the question that had the largest number of children expressing fear (about sleeping in a place where bats were able to fly around) and compared it with the question concerning whether children cared about bats. A total of 94 children provided valid responses to both questions. Within this group, 54 children (57%) indicated that they would feel some fear. Even among this group who indicated that they would feel fear, 72% still said that they cared about bats. This large overlap between these two groups indicates that a fear orientation toward bats existed alongside a caring orientation. In fact, based on a Pearson chi-square test, there was no significant difference in the percentage of caring between the children who indicated fear of bats and those who did not (72% vs. 85%, $p = 0.141$). Thus, based on these two questions analyzed, not only do fear and caring coexist, but children who indicated fear of bats were just as likely to care about bats as children who did not indicate fear of bats.

We also pursued two other questions: First, was fear or caring associated with attribution of thoughts and feelings to bats? The answer was no for fear, and yes for caring. Children who indicated that they cared about bats were more likely ($p < 0.005$, Fisher exact test) to believe that bats have thoughts (94%) and feelings (93%) than children who said that they did not care about bats (64% thoughts, 65% feelings). Second, were children who reported more fear of bats more likely to use anthropocentric justifications for keeping them in captivity? The answer was no, they were not.

Moral Basis for Keeping or Not Keeping Bats in Captivity

Toward first assessing the potential underpinnings of moral judgment, results showed that the majority of children said that bats in the exhibit have feelings (88%) and thoughts (88%). About half (48%) viewed the bats as capable of being blamed for bad behavior toward another bat, although such attributions decreased with age: 6–7-year-olds (63%), 9–10-year-olds (63%), 12–13-year-olds (43%), and 15–16-year-olds (23%) ($p = 0.001$). Similarly, about half (53%%) viewed the bats as capable of being blamed for bad behavior toward a person; and again such attributions decreased with age: 6–7-year-olds (73%), 9–10-year-olds (59%), 12–13-year-olds (48%), and 15–16-year-olds (31%) ($p = 0.001$). In contrast, virtually all (97%) of the children said that if a person mistreated a bat, the person should be blamed.

Eighty-six percent of the children believed that bats had rights, including the right to live (32%), and to their autonomy (25%). Developmentally, the older children asserted that bats had rights to a greater extent than the younger children (6–7, 58%; 9–10, 89%; 12–13, 89%; 15–16, 100%; $p = 0.0005$). When asked about specific rights, 97% of the children said that bats have the right not to be killed by humans, 95% said that bats have the right to be wild, and 93% said that bats have the right to live free.

We asked children to explain their reasons for why bats had the right not to be killed by humans, to be wild, and to live free. We coded their reasons by means of the typology summarized in Table 2. Table 3 reports the percentage of participants that used each category. As shown in Table 3, across all three questions the majority of children's justifications were biocentric (64%, 83%, and 78%, respectively) with use of anthropocentric reasoning only in the first question (16%, 0%, and 0% respectively). Developmental differences were found on the third question (why bats have the right to live free): Compared with the younger children, the older children provided more biocentric justifications (6–7, 33%; 9–10, 76%; 12–13, 89%; and 15–16, 93%; $p = 0.002$).

While the large majority of children asserted that bats had the right to be wild and to live free, it was also the case that 85% of the children said it was all right to keep bats in zoos, and

72% said that zoos did not violate the rights of bats by keeping them in the Australia house. Even among the children who said that bats have the right to be wild, only 5% said that it was not all right to keep bats in a zoo, and only 12% said that the zoo violated the rights of bats by keeping them in the Australia House. Similarly, of the children who said that bats have the right to live free, only 6% said that it was not all right to keep bats in a zoo, and 14% said that the zoo violated the rights of bats by keeping them in the Australia House.

The justification data go some initial distance to help explain how many of the children could maintain what might seem contradictory positions (e.g., that bats have the right to be wild and to live free, but that the zoo does not violate the rights of bats by keeping them in the Australia House). For this group of children, 27% reasoned that the environment within the zoo was congruent enough with the wild environment, such that there was no contradiction. For example, one child said "because if they recreate the environment so the bat thinks it is in the wild, then it's okay; but if you just put it in a little glass box then that's not right." Twenty-five percent said that the welfare of individual bats or of the species trumped the specific rights. For example, one child said "not really [it doesn't violate rights] because…they get taken care of, they get fed and everything." Another child said, "Yes, they have the right to live but like I said before the disease part; even though they might think it's not fair if they were to talk to us, it's really for their own good." Moreover, out of the children who said that bats have the right to be wild, 44% said that it was better for bats in the zoo, and 81% said that the zoo helps protect the bats from becoming extinct.

Discussion

Our results showed that most of the children brought forward a caring orientation toward bats. The majority of children, for example, said that they cared about bats, liked bats, would care if there were no bats in the world, and that it would matter to them personally if they lived their whole life without ever seeing a live bat, in the zoo or in the wild. That said, some of the children also indicated that they were somewhat scared of bats. For example, almost three quarters of the children said that they would prefer not to sleep in a place where bats were able to fly around, and about half said that in such a context they would have some fear. Moreover, we were able to establish that in the same children a fear orientation toward bats existed alongside of a caring orientation. Compared with the younger children, the older children were less fearful of bats, welcomed greater contact with bats, and said that they felt more alert while in the presence of bats. Taken together, these findings provide empirical support for the proposition that children can affiliate positively with—and bring a moral relationship to—an aspect of nature that they simultaneously fear.

One limitation of this study is that the bats in the exhibit were Rodrigues fruit bats, which have diminutive noses, large heads with conspicuous eyes, and other features that probably elicit less fear than many other species of bats. Thus, a future study could investigate similar issues but with another, more frightening, species of bats. In that context, an interesting question is whether children would care about the bats less than we found in this current study. More generally, future studies could investigate whether, and if so how, children care about other types of frightening animals (e.g., poisonous snakes and spiders).

Not only did the children we interviewed care about bats, but the large majority of children accorded bats the right to live free and to be wild. Yet most of the same children also said that it was all right to keep bats in a zoo, and that a zoo does not violate the rights of bats by keeping them in captivity. How could this be? There are three parts to our provisional answer. First, our results showed that while the majority of children accorded bats feelings and thoughts, only about half believed that bats could be morally responsible for bad behavior to either other bats or toward a person, while virtually all of the children said that if a person mistreated a bat, the person was morally responsible. Thus, it may be that when children accord bats (or other animals) rights, the rights are constrained in scope and considered less stringent as compared with human rights. Second, at least some of the children provided reasons that addressed the seeming contradiction. For example, some children said that the environment within the zoo was congruent enough with the environment in the wild (e.g., "like the wild and this is the same cause its got trees and water") such that the bats experienced little or no difference. Some children also said that the bats in the zoo were well taken care of, and were even better off in the zoo than in the wild (e.g., "if the bats got sick, they here might be able to take care of it but in the wild maybe they might not"). Third, some children appeared to simply hold seemingly contradictory positions: one that bats had rights to live free and to be wild, and another that the zoo did not violate those rights by keeping them in captivity. As one child said: "[It is both all right and not all right to keep bats in the zoo.] Yes because people learn that there are different types of bats in the world and different species; and no because I still believe that they should be free and in natural habitats."

All three answers may characterize how different types of children conceptualize these difficult issues. Nonetheless, the answers seem troubling to us, particularly given the ecological mission of many zoos today. According to Rabb (2004), a century ago the primary focus of zoos was to display "a wide variety of exotic animals solely for the recreation of the public" (p. 237). Today, however, zoos have become conservation centers, and among their responsibilities they "strive to help society achieve a more sustainable and harmonious relationship with nature…[by] contributing to the careful management of the earth's biological resources, which includes captive and wild animal populations and viable ecosystems, and…inspiring others to celebrate and conserve nature" (p. 237). With this mission laid out, it is a plus that so many children we interviewed at the zoo cared about bats insofar as such caring may go some distance to promoting conservation at large (Myers and Saunders 2002; Myers and Russell 2004). But if the same children really believe that the zoo environment is largely congruent with the wild environment, then a question for future research is whether such commitments bode poorly for a wider conservation ethic. Similarly, while it may be true that life is more hazardous to animals living in the wild compared with a zoo, such hazards are fundamental to what it means for an animal or ecosystem to be wild. If children miss this point in a zoo context, then here again it is worth investigating whether the zoo's mission will come up short.

Acknowledgements
Funding was provided by a generous grant from the Elizabeth Morse Genius Charitable Trust to Brookfield Zoo. We also extend our thanks to Andrej Birjulin for initial assistance with quantitative analyses; Todd Gieseke, Paul Walenga, Erik Garrett, Tatiana Garrett, and Mike Roman for assistance with qualitative analyses, coding, and data collection; and Janine Keca, Jill Siegel, and Cara Dinatale for assistance with data collection.

References
Ascione, F. R. 1993. Children who are cruel to animals: A review of research and implications for developmental psychopathology. *Anthrozoös* 6: 226–247.
Bixler, R. D. and Floyd, M. F. 1997. Nature is scary, disgusting, and uncomfortable. *Environment and Behavior* 29: 443–467.
Bowd, A. D. 1983. Children's fears of animals. *The Journal of Genetic Psychology* 142: 313–314.

Covert, A. M., Whirren, A. P., Keith, J. and Nelson, C. 1985. Pets, early adolescents, and families. *Marriage and Family Review* 8: 95–108.

Fleiss, J. L., Levin, B. and Paik, M. C. 2003. *Statistical Methods for Rates and Proportions.* 3rd edn. New York: John Wiley & Sons.

Helwig, C. C. 2006. Rights, civil liberties, and democracy across cultures. In *Handbook of Moral Development*, 185–210, ed. M. Killen and J. G. Smetana. Mahwah, NJ: Lawrence Erlbaum.

Howe, D., Kahn, Jr., P. H. and Friedman, B. 1996. Along the Rio Negro: Brazilian children's environmental views and values. *Developmental Psychology* 32: 979–987.

Kahn, P. H., Jr. 1999. *The Human Relationship with Nature: Development and Culture.* Cambridge, MA: MIT Press.

Kahn, P. H., Jr. 2006. Nature and moral development. In *Handbook of Moral Development,* 461–480, ed. M. Killen and J. G. Smetana. Mahwah, NJ: Lawrence Erlbaum.

Kahn, P. H., Jr. and Friedman, B. 1995. Environmental views and values of children in an inner-city Black community. *Child Development* 66: 1403–1417.

Kahn, P. H., Jr. and Lourenço, O. 2002. Water, air, fire, and earth—A developmental study in Portugal of environmental moral reasoning. *Environment and Behavior* 34: 405–430.

Kaltenborn, B. P., Bjerke, T., Nyahongo, J. W. and Williams, D. R. 2006. Animal preferences and acceptability of wildlife management actions around Serengeti National Park, Tanzania. *Biodiversity and Conservation* 15: 4633–4649.

Kellert, S. R. 1996. *The Value of Life.* Washington, DC: Island Press.

Kohlberg, L. 1984. *Essays in Moral Development: Vol. II. The Psychology of Moral Development.* San Francisco: Harper & Row.

Landis, J. and Koch, G. 1977. The measurement of observer agreement for categorical data. *Biometrics* 33: 159–174.

Lawrence, E. A. 1993. The sacred bee, the filthy pig, and the bat out of hell: Animal symbolism as cognitive biophilia. In *The Biophilia Hypothesis,* 301–341, ed. S. R. Kellert and E. O. Wilson. Washington, DC: Island Press.

Maurer, A. 1965. What children fear. *Journal of Genetic Psychology* 106: 265–277.

Melson, G. F. 2001. *Why the Wild Things Are: Animals in the Lives of Children.* Cambridge, MA: Harvard University Press.

Myers, O. E., Jr. 2007. *The Significance of Children and Animals: Social Development and Our Connection to Other Species.* Rev. edn. West Lafayette, IN: Purdue University Press.

Myers, O. E., Jr. and Russell, A. 2004. Human identity in relation to wild black bears: A natural-social ecology of subjective creatures. In *Identity and the Natural Environment,* 67–90, ed. S. Clayton and S. Opotow. Cambridge, MA: MIT Press.

Myers, O. E., Jr. and Saunders, C. D. 2002. Animals as links toward developing caring relationships with the natural world. In *Children and Nature: Psychological, Sociocultural, and Evolutionary Investigations,* 153–178, ed. P. H. Kahn, Jr. and S. R. Kellert. Cambridge, MA: MIT Press.

Nevers, P., Gebhard, U. and Billmann-Mahecha, E. 1997. Patterns of reasoning exhibited by children and adolescents in response to moral dilemmas involving plants, animals, and ecosystems. *Journal of Moral Education* 26: 169–186.

Piaget, J. 1960. *The Child's Conception of the World.* Totowa, New Jersey: Littlefield, Adams and Co. (Original work published 1929)

Rabb, G. B. 2004. The evolution of zoos from menageries to centers of conservation and caring. *Curator* 47: 237–246.

Rawls, J. 1971. *A Theory of Justice.* Cambridge: Harvard University Press.

Regan, T. 1983. *The Case for Animal Rights.* Berkeley: University of California Press.

Rolston, H., III. 1989. *Philosophy Gone Wild.* Buffalo, NY: Prometheus Books.

Scheffler, S. 1992. *Human Morality.* New York: Oxford University Press.

Thompson, B. 1996. AERA editorial policies regarding statistical significance testing: Three suggested reforms. *Educational Researcher* 25: 26–30.

Turiel, E. 1998. Moral development. In *Social, Emotional, and Personality Development*, 863–932, ed. W. Damon, *Handbook of Child Psychology.* 5th edn, Vol. 3: ed. N. Eisenberg. New York: Wiley.

Ulrich, R. S. 1993. Biophilia, biophobia, and natural landscapes. In *The Biophilia Hypothesis,* 73–137, ed. S. R. Kellert and E. O. Wilson. Washington, DC: Island Press.

Comparison of Vegetarians and Non-Vegetarians on Pet Attitude and Empathy

Brooke Dixon Preylo and Hiroko Arikawa

Forest Institute of Professional Psychology, Springfield, Missouri, USA

Address for correspondence:
Hiroko Arikawa,
2885 W. Battlefield Rd.,
Springfield, MO, 65807,
USA. E-mail:
hirokoar@gmail.com

ABSTRACT Past research found that positive attitudes toward animals are positively correlated with human-directed empathy. One of the most common reasons for becoming a vegetarian is to avoid cruelty toward animals. Based on the above literature, we hypothesized that vegetarians, especially moral vegetarians, would show higher human-directed empathy and more positive attitudes toward pets and other animals than non-vegetarians. Seventy-two vegetarians and 67 non-vegetarians participated in the study. Pet attitudes were measured using the modified Pet Attitude Scale (PAS-M), and human-directed empathy was measured with the Interpersonal Reactivity Index (IRI), which has four subscales. Vegetarian males had significantly higher empathy and significantly more positive attitudes toward pets compared with non-vegetarian males; however, there was no differences among females. There were no differences between moral vegetarians and non-moral vegetarians on human-directed empathy and attitude toward pets. Empathy toward humans and attitudes toward pets were positively correlated for both vegetarians and non-vegetarians. We conceptualized the dietary choice of a vegetarian as a lifestyle that can be explained by their political thinking, personality, and personal value systems.

Keywords: animal cruelty, diet, empathy, pet attitude, vegetarians

"Vegetarian" is a term that has many meanings and has been used to describe a wide variety of eating styles. Vegetarianism has been defined as a dietary style that is "characterized by the consumption of plant foods and the avoidance of some or all animal products" (Perry et al. 2001). According to a Harris Poll, about 2.3% of Americans (3% women, 2% men) never consume meat, poultry, and sea food, and consider themselves practicing vegetarians (Stahler 2006). Individuals choose a vegetarian lifestyle for a variety of reasons and the scientific literature offers diverse explanations. Cooper, Wise and Mann (1985) list "religious prohibitions, cultural beliefs, health benefits and counterculture attitude…" as reasons for following a vegetarian diet. In their study, examining the psychological and cognitive characteristics of vegetarians, 60% of the participants cited the "desire to avoid animal cruelty" as one reason for their vegetarianism. Beardsworth and Keil (1993) suggested that "the rearing, transporting and slaughter of food animals is challenged on moral grounds in that it is regarded as entailing unacceptable suffering or violation of animals' rights." Health implications and the morality of eating meat are two of the most commonly cited reasons for choosing a vegetarian diet. Many vegetarians are equally concerned about both topics. In fact, Amato and Partridge (1989) reported "concerns with health and morality are commonly presented as the two most influential factors in assuming a vegetarian food position." Rozin, Markwith and Stones (1997) found that 43.7% of their participants indicated that the healthiness of a meatless diet was a primary reason for becoming vegetarian, followed by the wastefulness of meat as food (38.2%), and moral reasons of killing (35%) and causing suffering to animals (35%).

It is important to understand empathy when examining prosocial behavior. Prosocial behavior is behavior that is positive in nature, such as sharing, volunteering, and helping. Counseling is a prosocial act (Carey, Fox and Spraggins 1988) and empathy is considered one of the most fundamental elements of counseling, according to Carl Rogers (1951). Empathy has also been linked with altruistic behavior (Davis 1983; Litvak-Miller and McDougall 1997; Unger and Thumuluri 1997). Daly and Morton (2003) pointed out that empathy has been used interchangeably with words like compassion, kindness, sympathy, and sentimentality. It has been referred to as sharing in another's perceived emotions or an affective reaction to another's distress. Davis (1983) suggested that there is more to empathy than feelings. He contended that it has both cognitive and affective components, in that it is a construct that has roots in emotion, as in sympathetic involvement, as well as a cognitive awareness and the ability to recognize another's feelings.

Recent literature suggests that empathy toward animals is positively related to human-directed empathy (Paul 2000). Ascione and Weber (1996) reported that children who had participated in a year-long educational program on humane attitudes had higher humane attitudes at one- and two-year follow-up than children in a control group. Furthermore, empathic attitudes toward animals of those children in the program showed evidence of generalizing to humans. The Antivivisection League, which protests the use of animals for scientific research, held a conference in 1998 to explore the relationship between violence towards animals and violence towards people. There it was pointed out that there is a close relationship between violence against animals, violence against children, and violence against women (Lockwood and Ascione 1998; Ascione and Arkow 1999). In one study of adolescent boys who were firesetters, it was found that they had higher levels of cruelty to animals than non-firesetters, and lower levels of empathy (Walsh, Lambie and Stewart 2004). In another study examining cruelty to animals in children, Dadds, Whiting and Hawes (2006) found that cruelty to animals and compromised levels of empathy were linked to conduct disorder, which is a childhood disorder characterized by aggressive behavior toward people or animals, destruction of property, deceitfulness, theft, and serious violation of social rules and the basic rights of others (American Psychiatric Association 2000).

The purpose of the current study was to make a direct comparison between vegetarians and non-vegetarians on their human-directed empathy and attitude to pets. Based on the literature, two hypotheses were generated. The first was that vegetarians, regardless of the reasons for their dietary choice, would have more positive attitudes toward pets and other animals and higher levels of empathy toward humans, compared with non-vegetarians. The second hypothesis was that among vegetarians, those who chose their diet for reasons of animal welfare would have more positive attitudes toward pets and higher levels of human empathy than vegetarians who did not choose their diet for reasons of animal welfare.

Methods
Participants
Participants were 139 adults (88 women and 51 men; mean age = 32.4, SD = 11.72) from Springfield, Missouri in the USA. All participants were solicited from local supermarkets that were either vegetarian-friendly or vegetarian-neutral. Vegetarian-friendly markets were those that marketed specifically, but not exclusively, to vegetarian consumers. Vegetarian-neutral supermarkets were those that marketed to the general public, with no specific emphasis on vegetarian food, although such stores generally carry limited amounts of meat-free alternatives. Adult shoppers were asked if they would like to participate in the study. They were also told that if they took part, they could sign up for a prize draw, to win a $25.00 gift certificate from the supermarket. After agreeing to participate, volunteers were then asked to take a seat at a small table where they gave informed consent. After consenting, participants were given a packet of questionnaires to complete and were asked to deposit the completed packet forms into a box labeled "Completed Surveys." Upon completion, participants were asked if they would like to sign up for the gift certificate prize draw, as previously mentioned. Only those who wanted to sign up for the gift certificate were given a paper on which to provide their name, telephone number, and e-mail address. Participants placed completed prize-draw papers in a box separate from the one used for the questionnaires, in order to maintain anonymity of the responses. Upon completion of the prize draw, all papers containing personal information were shredded and discarded. All participants were treated in accordance with the approved protocol of the Institutional Review Board of Forest Institute of Professional Psychology.

Instruments
Two scales were used in addition to a demographic questionnaire. The Pet Attitude Scale-Modified (PAS-M) (Munsell et al. 2004), which is a modified version of the Pet Attitude Scale (PAS) (Templer 1981), was used to measure participants' attitudes toward pets. The PAS-M was chosen as an instrument for the current study because it is the most frequently used scale to measure attitudes toward pet animals, therefore has good construct validity. In addition, the PAS-M measures not only attitudes toward pets but also contain a few items to examine attitudes to animals in general. The PAS-M has 18 items and utilizes a Likert-type format (1 = strongly disagree, 2 = moderately disagree, 3 = slightly disagree, 4 = unsure, 5 = slightly agree, 6 = moderately agree, 7 = strongly agree) to assess attitudes toward pets. The total score ranges from 18 to 126—the higher the score, the more positive the attitude toward pets. Examples of the items include "House pets add happiness to my life," " I hate animals," and "You should treat house pets with as much respect as you would a human member of your family." Test-retest reliability is 0.92 and Cronbach's alpha is 0.93. Meaningful correlations have been obtained between the PAS-M and the Allport-Vernon Lindsey Study of Values in addition to the Personality Research Form. There are three factors in the PAS-M: Love and Interaction, Pets in the Home and Joy of Pet Ownership.

The Interpersonal Reactivity Index (IRI) was used (Davis 1983) as a measure of empathy because its assesses both cognitive and emotional domains. The IRI has 28 items, with responses given on five-point Likert-type scales (1 = does not describe me very well, 5 = describes me very well), and total scores range from 28 to140. It assesses four factors/subscales—two cognitive and two emotional—each of which consists of seven items. Subscales include Perspective Taking (PT) (e.g., "I try to look at everybody's side of a disagreement before I make a decision"), Fantasy (FS) (e.g., "I daydream and fantasize, with some regularity, about things that might happen to me"), Empathic Concern (EC) (e.g., "Sometimes I don't feel sorry for other people when they are having problems"), and Personal Distress (PD) (e.g., "When I see someone who badly needs help in an emergency, I go to pieces"). Davis suggests that PT measures spontaneous tendencies of the test-taker to adopt the psychological perspective of another person or to entertain another's point of view. Of the four subscales, this one most clearly measures cognitive aspects of empathy. FS assesses the test-taker's ability to imagine himself or herself in place of fictional characters such as those found in movies and books. The FS scale is a measure of emotional empathy. EC examines the respondent's tendency to experience warmth, compassion, and concern for others. EC is the second measure of emotional empathy. And lastly, PD examines feelings of anxiety and distress in crisis situations. This is the second cognitive measure of empathy. Internal reliability of the IRI has been measured to be between 0.71 and 0.77, and test-retest reliability ranges from 0.61 to 0.71 (Davis 1983).

The demographic questionnaire asked all participants for their age and gender and whether or not they were vegetarian and pet owners. Individuals who were vegetarian were asked which type they were: lacto-ovo, lacto, ovo, vegan, fruitarian, or other. In addition, vegetarians were asked to indicate the reasons why they chose a vegetarian dietary style (health reasons, religious beliefs, political beliefs, animal rights/welfare, other), in order to differentiate between those who did so for animal welfare reasons and those who did not.

Construction of Groups

For the current study, a vegetarian was defined as a person who never ate meat, poultry, fish, or other seafood. Seventy-two participants (15 male, 57 female, 44 pet owners, 28 non-pet owners) identified themselves as vegetarian and 67 (36 male, 31 female, 32 pet owners and 35 non-pet owners) identified themselves as non-vegetarian. Of the vegetarians, 31 were lacto-ovo, 23 were lacto, five were ovo, 12 were vegans, and one was fruitarian.

The vegetarian group was further divided into two groups (animal welfare vs. other), based on the reasons given for choosing their dietary style. Regardless of the number of reasons given, vegetarians were classified as "animal welfare vegetarians" as long as animal welfare was one of the reasons given. Forty-nine (68.1%) of the vegetarians were categorized as animal welfare vegetarians.

Results

Table 1 presents the Pearson's correlation of the four subscales of the IRI and the PAS-M. Positive correlations ($p < 0.001$) were found within each of the four subscales of the IRI, and between the PAS-M and each subscale of the IRI. Table 2 shows the correlations of the four IRI subscales and the PAS-M for vegetarians ($n = 72$) and non-vegetarians ($n = 67$) separately. Table 3 shows the MANOVA F value indicating that vegetarians scored significantly higher than non-vegetarians on the EC, FS, and PT subscales and PAS-M ($p < 0.001$), and the PD subscale ($p < 0.01$).

Table 1. Intercorrelations between IRI subscale scores and PAS-M total scores.

	1	2	3	4	5
1 Empathic Concern	—	0.74*	0.83*	0.34*	0.64*
2 Fantasy		—	0.67*	0.58*	0.68*
3 Perspective Taking			—	0.38*	0.50*
4 Personal Distress				—	0.44*
5 PAS-M Total					—

* $p < 0.001$

Table 2. Intercorrelations between IRI subscale scores and PAS-M total scores by dietary style (vegetarians and non-vegetarians).

	1	2	3	4	5
Vegetarians (n = 72)					
1 Empathic Concern	—	0.59*	0.75*	0.02	0.35*
2 Fantasy		—	0.56*	0.38*	0.37*
3 Perspective Taking			—	0.22	0.23
4 Personal Distress				—	0.17
5 PAS-M Total					—
Non-Vegetarians (n = 67)					
1 Empathic Concern	—	0.78*	0.86*	0.60*	0.72*
2 Fantasy		—	0.68*	0.76*	0.76*
3 Perspective Taking			—	0.46*	0.57*
4 Personal Distress				—	0.64*
5 PAS-M Total					—

* $p < 0.001$

Table 3. Comparison of vegetarians versus non-vegetarians on empathy (IRI subscales) and pet attitude (PAS-M total scores).

	Vegetarians (n = 72)		Non-Vegetarians (n = 67)		F
	Mean	SD	Mean	SD	
Empathic Concern	20.61	4.87	16.16	5.82	23.96**
Fantasy	17.53	5.24	12.75	7.31	19.85**
Perspective Taking	19.15	4.70	15.36	4.92	21.62**
Personal Distress	12.94	7.30	9.7	5.50	8.80*
PAS-M Total	99.32	17.64	81.39	26.53	22.30**

* $p < 0.01$; ** $p < 0.001$

Since women tend to report higher empathy toward humans than men, as found in previous research (Lennon and Eisenberg 1989[7]), and pet owners tend to report more empathy toward animals than non-pet owners (Paul 2000), the variables of gender and pet ownership may be confounding the current findings. In order to examine the possible confounds of gender on empathy scores and pet ownership on PAS-M score, further analyses compared vegetarians and non-vegetarians within each gender on the IRI, and pet ownership status on

PAS-M. With regards gender, female participants scored significantly higher than males on the EC ($F = 35.92$, $p < 0.001$), FS ($F = 24.87$, $p < 0.001$), PT ($F = 22.21$, $p < 0.001$), and PD ($F = 17.41$, $p < 0.001$) subscales of the IRI and on the PAS-M ($F = 23.28$, $p < 0.001$). Within each gender, vegetarian males scored significantly higher than non-vegetarian males on the PAS-M, EC, FS, PT ($p < 0.01$), and PD ($p < 0.05$). However, there was no difference between vegetarians and non-vegetarians on the IRI subscales and the PAS-M among females (Table 4). There was no significant difference between vegetarians and non-vegetarians on the PAS-M among pet owners ($t = 0.295$). However, vegetarians scored significantly higher ($t = 6.757$ $p < 0.001$) on the PAS-M than non-vegetarians, among non-pet owners.

Tale 4. Comparison of vegetarians versus non-vegetarians on empathy (IRI subscale scores) and pet attitude (PAS-M total scores) by gender.

	Males (n =51)				
	Vegetarians (n = 15)		Non-Vegetarians (n = 36)		
	Mean	SD	Mean	SD	F
Empathic Concern	18.73	4.78	13.47	5.06	11.83**
Fantasy	16.20	4.72	9.92	7.21	9.61**
Perspective Taking	17.93	4.22	13.50	4.78	9.71**
Personal Distress	11.20	5.54	7.28	5.34	5.59*
PAS-M Total	95.20	18.67	71.81	23.31	11.88**
	Females (n = 88)				
	Vegetarians (n = 57)		Non-Vegetarians (n = 31)		
	Mean	SD	Mean	SD	F
Empathic Concern	21.11	4.82	19.29	5.09	2.74
Fantasy	17.88	5.35	16.03	6.02	2.19
Perspective Taking	19.47	4.80	17.52	4.19	3.64
Personal Distress	13.40	7.68	12.45	4.30	0.41
PAS-M Total	100.40	17.37	92.52	26.01	2.89

* $p < 0.05$; ** $p < 0.01$

The 76 pet owners in our sample showed significantly higher scores (empathy) on the EC ($F = 7.1$, $p < 0.001$), FS ($F = 20.89$, $p < 0.001$), PT ($F = 4.65$, $p < 0.05$), and PD ($F = 8.8$, $p < 0.01$) subscales of the IRI and significantly higher scores on the PAS-M (more positive attitudes toward pets) ($F = 67.61$, $p < 0.001$) than the 63 non-pet owners. The same comparisons were made within vegetarians and non-vegetarians, in order to examine the possible confound of vegetarian status. It was found that non-vegetarian pet owners scored significantly higher ($p < 0.001$) on the EC ($F = 24.21$), FS ($F = 31.74$), PT ($F = 14.37$), and PD ($F = 14.27$) subscales of the IRI, and significantly more positive attitudes toward pets ($F = 93.97$, $p < 0.001$) than non-pet owning non-vegetarians. Among vegetarians, there was no significant difference between pet owners and non-pet owners on the empathy scales. However, pet owners showed significantly more positive attitudes toward pets than non pet-owners ($F = 7.02$, $p < 0.05$).

Animal welfare vegetarians ($n = 49$) and non-animal welfare vegetarians ($n = 23$) were compared on their levels of empathy and pet attitude. There was no difference on any the subscales of the IRI; EC ($F = 0.53$), FS ($F = 0.02$), PT ($F = 0.04$), PD ($F = 0.01$) or on PAS-M scores ($F = 0.02$).

A discriminant function analysis was conducted to determine whether the four subscales of the IRI and the PAS-M could predict dietary choice. The function significantly ($\Lambda = 0.811$, $\chi^2 = 28.19$, $df = 5$, $n = 139$, $p < 0.001$) predicted differences between vegetarians and non-vegetarians. Function coefficients indicate that EC, PAS-M and PT were the strongest predictors of vegetarian diet. Overall, 68.3% of the sample was correctly classified. Vegetarians had a function mean of 0.46 while non-vegetarians had a mean of –0.5. The results suggest that people with higher levels of empathic concern, more ability to take the perspective of others, and more positive attitudes toward pets are more likely to be vegetarian than not.

Discussion

Compared with the non-vegetarian participants, the vegetarians in our study had more empathic concerns and were more likely to be able to consider another person's perspective. Additionally, they were more likely to be able to visualize themselves in the position of others and were more likely to experience distress when another person was in a difficult situation. They also had a more positive attitude toward pets. However, theses differences were only found amongst men—there was no difference in empathy and pet attitude between female vegetarians and female non-vegetarians. This may be due to the influence on men of media-portrayed characteristics of masculinity, in which men are preoccupied with red meat and are less likely to seek healthier meals (Gough 2007). Also, males are more likely to hunt game animals and to fish, and then barbeque them at gatherings of family and friends. It is possible, though, that males—despite their tough image—who choose a vegetarian diet are more sensitive and empathic toward others than meat-eating males.

Contrary to expectation, animal welfare reasons for being a vegetarian did not differentiate levels of empathy among vegetarians. In fact, scores were almost identical. The same was true even for attitudes toward pets. This may possibly be attributed to participants who endorse animal rights/extremist views. For example, some individuals who belong to animal rights organizations do not believe in owning pets. In fact, in extreme cases, pet ownership may be viewed as a form of mistreatment of animals. Therefore, when endorsing items on the PAS-M, they might disagree with an item such as #3, "I would like to have a pet in my home" while strongly disagreeing with an item such as #17 "I hate animals." While these individuals may have a great deal of concern for animal welfare, their PAS-M scores would not reflect that, since it is primarily a measure of pet attitude, and only a few items focus on animal attitudes in general, resulting in inconsistent endorsement of items and, possibly, low total scores.

In general, we found that individuals with more positive pet attitudes were more empathic toward people. This result was even stronger among non-vegetarians than vegetarians. These results were unexpected and again may be the result of individuals who do not support having animals as pets. Alternatively, vegetarians in general may have more heterogeneity among them and so do not show strong patterns of personality when examined as a group. Nevertheless, in our current sample, our general hypothesis that people who have more positive attitudes toward animals are more empathic toward people was sustained, regardless of dietary choice.

There are some exceptions in which empathy toward animals does not extend to empathy toward humans. Probably the most notorious example of a historical figure who was concerned with animal rights and was also a vegetarian was Adolf Hitler. Despite his beliefs and dietary style, he also exhibited extreme cruelty to innocent people, including children (Arluke and Sax 1992). Certainly, then, it is possible that some individuals feel more empathic toward animals than humans. A further example of this is individuals who shame or humiliate people in public in order to protest about various aspects of animal treatment—in more extreme cases, people are physically harmed in order to protect animals. In the Edo period of Japan, Tsunayoshi Tokugawa, the fifth shogun and vegetarian Buddhist, earned an insulting nickname of Dog Shogun after issuing the notorious *Laws of Compassion*, an extreme protection of all animals, especially dogs. The roaming dogs cluttered streets, and those animals that were sick or injured were housed and fed while those who harmed animals were executed (Bodart-Bailey 2007). This behavior may be partially explained by past research which compared moral and health vegetarians (Rozin, Markwith and Stoess 1997). They found that moral vegetarians have stronger tendencies to believe that meat-eaters have less desirable personalities, are more aggressive, and are more animal-like. Hitler also considered Jews to be a group of people who lacked morality and compassion, especially in their treatments of animals in scientific research, and considered them as less desirable than most animals (Arluke and Sax 1992)

Seven participants in our study gave reasons for their responses to items on either the PAS-M or the IRI, or both, even though neither of the questionnaires invited participants to write comments on them. For example, one individual explained his low score on item #2 of the PAS-M ("My pet means more to me than any of my friends [or would if I had one]"), stating that "it is unfair to treat one group better than the other." There were several participants who gave low scores, such as the previous example, but then gave explanations for their choice, indicating that perhaps the question itself was stated in a way that they disliked, rather than them simply disliking people or animals. Interestingly, all the individuals who wrote comments were vegetarians. It is likely that they felt a strong need to express their opinions on the issues raised because their unique lifestyle as a vegetarian is by conscious choice, supported by often very strong personal belief systems. This is in contrast to the huge proportion (97%) of people who eat meat in the US, people who are probably rarely asked why they consume meat.

It is reasonable to state that for many vegetarian individuals, especially for moral vegetarians, being a vegetarian is not merely a dietary preference, but it is their personal statement. Hamilton (2006) conducted in-depth interviews of vegetarians and meat-eaters and found that moral vegetarians' refusal to consume meat "symbolizes complicity in what is immoral," although they are aware that their protest does not affect the meat market.

Future investigations should include closer scrutiny of heterogeneity among vegetarians, gender differences, and psychological variables. The members of animal rights organizations would appear to be appropriate participants for such research. Measuring masculinity/femininity would be useful, as it may very well be related to empathy toward humans and attitudes toward animals.

Acknowledgements
We gratefully acknowledge the support extended by Dr. Adrian J. Whitmire, Dr. Frances Parks, Ms. Shelly Vaugine and Dr. Partick C. Gariety. We wish to thank two anonymous reviewers and Dr. Anthony Podberscek for their helpful comments and recommendations for revision.

References
Amato, P. R. and Partridge, S. A. 1989. *The New Vegetarians: Promoting Health and Protecting Life.* New York: Plenum Press.
American Psychiatric Association. 2000. *Diagnostic and Statistical Manual for Mental Disorders.* 4th edn. Text Revision. Washington, DC: Author.
Arluke, A. and Sax, B. 1992. Understanding Nazi animal protection and the Holocaust. *Anthrozoös* 5: 6–31.

Ascione, F. R. and Arkow, P. 1999. *Child Abuse, Domestic Violence, and Animal Abuse.* West Lafayette, IN: Purdue University Press.

Ascione, F. R. and Weber, C. V. 1996. Children's attitudes about the humane treatment of animals and empathy: One-year follow up of a school-based intervention. *Anthrozoös* 9: 188–195.

Beardsworth, A. D. and Keil, E. T. 1993. Contemporary vegetarianism in the UK: Challenge and incorporation? *Appetite* 20: 229–234.

Bodart-Bailey, B. M. 2007. *The Dog Shogun: The Personality And Policies of Tokugawa Tsunayoshi.* Hawaii: University of Hawaii Press.

Carey, J. C., Fox, E. A. and Spraggins, E. F. 1988. Replication of structure findings regarding the Interpersonal Reactivity Index. *Measurement and Evaluation in Counseling and Development* 21: 102–105.

Cooper, C. K., Wise, T. N. and Mann, L. S. 1985. Psychological and cognitive characteristics of vegetarians. *Psychosomatics: Journal of Consultation Liaison Psychiatry* 26: 521–527.

Dadds, M. R., Whiting, C. and Hawes, D. J. 2006. Associations among cruelty to animals, family conflict and psychopathic traits in childhood. *Journal of Interpersonal Violence* 21: 411–429.

Daly, B. and Morton, L. L. 2003. Children with pets do not show higher empathy: A challenge to current views. *Anthrozoös* 16: 298–314.

Davis, M. H. 1983. The effects of dispositional empathy on emotional reactions and helping: A multidimensional approach. *Journal of Personality* 51: 167–184.

Gough, B. 2007. Real men don't diet: An analysis of contemporary newspaper representations of men, food and health. *Social Science & Medicine* 64: 326–337.

Hamilton, M. 2006. Eating death. *Food, Culture and Society* 9: 157–177.

Lennon, R. and Eisenberg, N. 1987. Gender and age differences in empathy and sympathy. In *Empathy and Its Development*, 195–217, ed. N. Eisenberg and J. Strayer. New York: Cambridge University Press.

Litvak-Miller, W. and McDougall, D. 1997. The structure of empathy during middle childhood and its relationship to prosocial behavior. *Genetic, Social & General Psychology Monographs* 123: 303–321.

Lockwood, R. and Ascione, F. R. 1998. *Cruelty to Animals and Interpersonal Violence.* West Lafayette, IN: Purdue University Press.

Munsell, K. L., Canfield, M., Templer, D. I., Tangen, K. and Arikawa, H. 2004. Modified pet attitude scale. *Society & Animals* 12: 137–142.

Paul, E. S. 2000. Empathy with animals and with humans: are they linked? *Anthrozoös* 13: 194–202.

Perry, C. L., McGuire, M. T., Neumark-Sztainer, D. and Story, M. 2001. Characteristics of vegetarian adolescents in a multiethnic urban population. *Journal of Adolescent Health* 29: 406–416.

Rogers, C. R. 1951. *Client-Centered Counselling.* Boston: Houghton-Mifflin.

Rozin, P., Markwith, M. and Stoess, C. 1997. Moralization and becoming a vegetarian: the transformation of preferences into values and the recruitment of disgust. *Psychological Science* 8: 67–73.

Stahler, C. 2006. How many adults are vegetarians? <www.wrg.org/journal/vj2006issue4/vj2006issue4poll.htm> Accessed September 6, 2007.

Templer, D. I., Salter, C. A., Dickey, S., Baldwin, R. and Veleber, D. M. 1981. The construction of a Pet Attitude Scale. *The Psychological Record* 31: 343–348.

Unger, L. S. and Thumuluri, L. K. 1997. Trait empathy and continuous helping: The case of voluntarism. *Journal of Social Behavior and Personality* 12: 785–800.

Walsh, D., Lambie, I. and Stewart, M. 2004. Sparking up: Family, behavioral and empathy factors in adolescent firesetters. *American Journal of Forensic Psychology* 22: 5–32.

THE ANIMALS READER
The Essential Classic and Contemporary Writings

Edited by Linda Kalof and Amy Fitzgerald

"...from Aristotle to postmodern philosophers...

...from orangutans to cyborgs...

The Animals Reader presents a wonderful diversity of perspectives on animals and, in consequence, ourselves."

Boria Sax, author of *Crow*, *Animals in the Third Reich* and *The Mythical Zoo*

The study of animals – and the relationship between humans and other animals – is now one of the most fiercely debated topics in contemporary science and culture. As the first book of its kind, *The Animals Reader* provides a framework for understanding the current state of the multidisciplinary field of animal studies. This anthology will be invaluable for students across the Humanities and Social Sciences as well as for general readers.

Contributors Include:

Aristotle • Steve Baker • Marc Bekoff • Jeremy Bentham • John Berger • Jonathan Burt
Gilles Deleuze • Descartes • Donna Haraway • Lévi-Strauss • Randy Malamud • Steven Mithen
Michel de Montaigne • Martha Nussbaum • Pliny • Plutarch • Tom Regan • Harriet Ritvo
Boria Sax • Peter Singer • Marjorie Spiegel • Yi-Fu Tuan • Sarah Whatmore

February 2007 • 448pp • 10 b/w illus
ISBN 978 1 84520 470 9 (PB) £19.99 • $34.95
ISBN 978 1 84520 469 3 (HB) £60.00 • $105.00
www.bergpublishers.com

NEWS & ANALYSIS

Section on Animal–Human Interaction: Research & Practice (AHI)

The Section on Animal–Human Interaction: Research & Practice (AHI) was approved for full section status by the Executive Board of the Society of Counseling Psychology at the International Counseling Psychology Conference this year. Areas of interest to the AHI include the role of animals in empathy development, human health, animal-assisted therapies, the ability to form and express attachments, reaction to grief and loss, the challenge of aging, and other developmental passages throughout the lifespan. For more information, please contact Mary Lou Randour, Chair: Randour@comcast.net

ISAZ/WALTHAM Research Award

The International Society for Anthrozoology (ISAZ) Board is thrilled to announce the launch of the ISAZ/WALTHAM Collaborative Research Award. The purpose of this collaboration is to stimulate new research in the area of human–animal interactions, with particular interest in the role of pets in the lives of elders, pets enhancing healthy longevity, and understanding the barriers to pet ownership. A total of US$22,000 in grant funds will be available for distribution. One or more projects may be funded within this budget. The deadline to submit completed applications is: **January 1, 2009**. The grant application is now available for download from the ISAZ homepage (www.isaz.net).

New Books

Taxidermic Signs: Reconstructing Aboriginality

By Pauline Wakeham

Pauline Wakeham decodes the practice of taxidermy as it was performed in North America from the late nineteenth century to the present, revealing its connection to ecological and racial discourses integral to the maintenance of colonial power. Moving beyond the literal practice of stuffing skins, Wakeham theorizes taxidermy as a sign system that conflates "animality" and "aboriginality" within colonial narratives of extinction. Through a series of provocative case studies, Wakeham demonstrates how the semiotics of taxidermy travel across diverse cultural texts. From the display of animal specimens and aboriginal artifacts in the Banff Park Museum, to the ethnographic films of Edward S. Curtis and Marius Barbeau, to the fetishization of aboriginal remains in the Kennewick Man and Kwäday Dän Ts'inchi repatriation cases, Wakeham argues that taxidermy's sign system reinvents mythologies of disappearing wildlife and vanishing Indians while simultaneously valorizing the power of Western technologies to memorialize these figures.

Seeking to destabilize the hierarchies of anthropocentric white supremacy, Wakeham presents an analysis of taxidermy as both a material practice and a symbolic system foundational to colonial authority in North America and still vital to the maintenance of power asymmetries today. Published in 2008 by University of Minnesota Press. ISBN-13: 978-0816650552 (paperback); ISBN-13: 978-0816650545 (hardback).

Wings in the Desert: A Folk Ornithology of the Northern Pimans
By Amadeo M. Rea

There is a common but often unspoken arrogance on the part of outside observers that folk science and traditional knowledge—the type developed by Native communities and tribal groups—is inferior to the "formal science" practiced by Westerners. In this lucidly written and humanistic account of the O'odham tribes of Arizona and Northwest Mexico, ethnobiologist Amadeo M. Rea exposes the limitations of this assumption by exploring the rich ornithology that these tribes have generated about the birds that are native to their region. He shows how these peoples' observational knowledge provides insights into the behaviors, mating habits, migratory patterns, and distribution of local bird species, and he uncovers the various ways that this knowledge is incorporated into the communities' traditions and esoteric belief systems. Drawing on more than four decades of field and textual research along with hundreds of interviews with tribe members, Rea identifies how birds are incorporated, both symbolically and practically, into Piman legends, songs, art, religion, and ceremonies. Through highly detailed descriptions and accounts loaded with Native voice, this book is the definitive study of folk ornithology. It also provides valuable data for scholars of linguistics and North American Native studies, and it makes a significant contribution to our understanding of how humans make sense of their world. It will be of interest to historians of science, anthropologists, and scholars of indigenous cultures and folk taxonomy. Published in 2007 by The University of Arizona Press. ISBN: 978-0-8165-2459-4 (hardback).

Conferences

Experiential Learning in Humane Education: Involve Me and I Will Understand
April 24 to 25, 2009

This conference, hosted by The American Society for the Prevention of Cruelty to Animals (ASPCA) and Green Chimneys, will be held at **Green Chimneys, Brewster, New York, USA**. It will explore how experiential learning promotes empathetic responses and will look at the role living animals play establishing a healthy relationship, in promoting empathy, and in bringing about responsible animal stewardship. For more information, please contact Michael Kaufmann, Green Chimneys Farm and Wildlife Director (mkaufmann@greenchimneys.org) or Sheryl L. Pipe, Ph.D., Senior Director, Humane Education, ASPCA (sherylp@aspca.org), or visit: www.greenchimneys.org.

News & Analysis

Animal Behavior Society Annual Meeting
June 22 to 26, 2009

The 46th Annual Animal Behavior Meeting will be held in **Pirenópolis, Brazil**. For further information, send an e-mail to: rhfmacedo@unb.br or view the official website: www.animal-behavior.org/Brazil09

43rd Congress of the International Society for Applied Ethology
July 6 to 10, 2009

This congress will be held at the **Cairns Convention Centre, Cairns, Queensland, Australia**. The theme is Applied Ethology for Contemporary Animal Issues, and the sub-themes are:

- animal welfare assessment and enhancement
- management of unwanted animals
- animal emotion and cognition
- animals in extensive and natural environments
- animal–human interactions

For more information, send an e-mail to: isae2009@icms.com.au or view the official website: www.isae2009.com

International Academic and Community Conference on Animals and Society
Minding Animals

July 13 to 18, 2009

The University of Newcastle and the Animals and Society (Australia) Study Group are hosting this conference in **Newcastle, Australia**. It will bring together a broad range of academic disciplines and representatives from universities, non-government organizations and the community, industry, and governments from around the world. Conference delegates will examine the interrelationships between human and nonhuman animals from cultural, historical, geographical, environmental, moral, legal, and political perspectives.

The conference will have six major themes and objectives:

- To reassess the relationship between the animal and environmental movements in light of climate change and other jointly-held threats and concerns
- To examine how humans identify and represent nonhuman animals in art, literature, music, science, and in the media and on film
- How, throughout history, the objectification of nonhuman animals and nature in science and society, religion, and philosophy, has led to the abuse of nonhuman animals and how this has since been interpreted and evaluated
- To examine how the lives of humans and companion and domesticated nonhuman animals are intertwined, and how science, human medicine, and veterinary medicine utilize these important connections

News & Analysis

- How the study of animals and society can better inform both the scientific study of animals and community activism and advocacy
- How science and community activism and advocacy can inform the study of non-human animals and society

Plenary speakers include: Carol Adams, Marc Bekoff, Dale Jamieson, Tom Regan, Andrew Rowan, James Serpell, Peter Singer, and Jennifer Wolch. Registration and abstract submission is now open online at: www.mindinganimals.com. If you have any queries regarding the conference, please send an e-mail to: mindinganimals@pco.com.au

ISAZ/HAI International Conferences
Human–Animal Interaction: Impacting Multiple Species
October 21 to 25, 2009

This is made up of two sequential conferences—the 18th Annual International Society for Anthrozoology conference (October 21 to 23) and the 1st Annual Conference of the Research Center for Human–Animal Interaction (October 23 to 25)—which will be held in **Kansas City, Missouri, USA**. Here, researchers and practitioners will be able to share their latest findings and program outcomes, For further information, go to either www.isaz.net or www.rechai.missouri.edu

BOOK REVIEWS

Bears: A Brief History

Bernd Brunner, translated from the German by Lori Lantz. New Haven: Yale University Press, 2007. 259 pages. ISBN: 978-0-300-12299-2 (hardback)

Reviewed by: Juliet Clutton-Brock. E-mail: juliet.cb@btinternet.com

This small book about bears and their relationships with humans has similarities with the volume named *Bear* by Robert E. Bieder in the series *Animal*, published by Reaktion Books (2005). The authors may think this is unfortunate but it can also be seen as a benefit for, although the two books follow much the same pattern, their approach is different, and both are equally fascinating. Both books are full of intriguing legends, traditions, and anecdotes, but *Bears* has all the added charm and empathy of the author's earlier work, *The Ocean at Home: An Illustrated History of the Aquarium* (2003, reviewed in *Anthrozoös* 2006, 19(1): 89–90).

Like *Bear*, Brunner's book begins with a short description of the eight living species of bears and tracks their fossil history from the small, ancestral, bear-like animal of around 22 million years ago. The second chapter in *Bears* introduces the main theme of the book, which is the kinship between humans and bears that has persisted through history and is represented in the innumerable myths and traditions that prevail in many parts of the world. This is followed by "The Mystery of the Cave Bear" in which Brunner describes how finds of the remains of bears in Pleistocene cave deposits have led to beliefs in a bear cult between the Ice Age hunters and the giant brown bears that probably hibernated in the caves. Archaeologists have disparaged these notions of a relationship other than that of hunter and prey, but in describing the remains of bears in the 35,000-year-old cave of Chauvet in the Ardèche, France, Brunner fails to mention the detail of the bear's skull found perched on a slab of rock, together with grains of charcoal. Surely there was some ritual significance in the purposeful placing of this skull by the hunters and artists, whose bare footprints have been found in the cave, and it must indeed have seemed that bears possessed remarkable powers to bring them back to life after months of long winter hibernation under ground.

Although a pragmatic attitude is taken to the often presumed but unknown relationships between bears and humans in the Ice Age, the strengths of this book lie in its fascinating descriptions of the rituals that have accompanied the hunting of bears by tribal peoples who have almost all believed that the soul of the bear must be appeased in death. These beliefs may be compared with the ruthless attitudes of the big game hunters of Victorian and later times, who shot for "sport" and for trophies all eight species of bear, including the giant panda (often classified by zoologists in a separate family, the Ailuropodidae). The killing of a panda by the sons of President "Teddy" Roosevelt is described thus: "He looked sleepily from side to side and he sauntered forth and walked slowly away into the bamboos. As soon as Ted came up we fired simultaneously at the outline of the disappearing panda. Both shots took effect. He was a splendid old male…"

Probably the best-known tradition concerning the raising of young bear cubs, which could include suckling by women and then their ritual killing when adult, is that of the Ainu people of Hokkaido in Japan. For them, bears were seen as gods whose spirits returned to the mountains after their killing had been appeased in a great annual festival.

Many such traditions are described by Brunner, from the beliefs and rituals that surround the hunting of polar bears by the Inuit to the description by the famous taxonomist Linnaeus of a bear hunt in Lapland to the Cree hunters of north eastern Canada who believed that the bears themselves would choose to be killed. The hunter would put a lit pipe into the dead bear's mouth and blow into it to fill the bear's throat to calm its spirit and prevent it from seeking revenge.

Chapter 14 is on "Bears on Show" and describes the many ways in which bears have been exploited since Roman times as the cruel victims of baiting and fighting "sports," as well as trained animals in circuses and street "dancing." Brunner describes one bizarre incident in which the Norwegian explorer, Roald Amundsen, believed that polar bears could be trained to pull sleds on an Arctic expedition. Carl Hagenbeck and animal trainer Reuben Castang apparently succeeded in training twenty-one polar bears to pull the sleds in small groups. However, the plan came to nothing because Castang refused to travel to the Arctic with the bears.

The sixteen chapters in the book are embellished with numerous, nineteenth century, black and white engravings, drawings, and photos that bring to life the close and often horribly cruel relationship that has endured between bears and their human hunters and captors for thousands of years. Fortunately, however, there are still species of bears found wild in many parts of the world, except in Africa where there has never been any indigenous species.

Readers may choose one of the two books to read, either Brunner's *Bears* or *Bear* in the *Animal* series, but what is surprising is that there are not a great many more books on this fascinating but endangered group of large mammals. Brunner's book ends with an Epilogue and a Bibliographic Essay that provides web addresses as well as a Selected Bibliography and an Index, but it should also have included WSPA (the World Society for the Protection Animals), which has specialized in trying to ban the keeping of dancing bears.

Dog Behaviour, Evolution and Cognition

Ádám Miklósi. Oxford University Press, 2007. 241 pages. ISBN: 978-0-19-929585-2 (hardback)

Reviewed by: Claire L. Corridan, Animal Behaviour, Cognition & Welfare Group, University of Lincoln, UK. E-mail: CCorridan@lincoln.ac.uk

Miklósi introduces this book by explaining how in 1994 he first started to focus his research on dog–human social interactions and was amazed that at that time, literature on the subject was simply non-existent. His aim was, and is, to provide an evolutionary framework that hypothesizes behavioral convergence between the two species. The fact that this book covers so many of the crucial aspects of dog ethology, genetics, behavior, and interactions with humans in such great detail is a testament to its author and the scientists cited throughout the book who have contributed to expansion of our knowledge in this field. Miklósi has also identified several areas where knowledge is lacking or minimal and how misinterpretation of the few studies that have been completed can lead to dramatic changes in breeding, dog selection, puppy socialization, training, and even policy and legislation. "We know, in terms of scientific validated knowledge, much less about dogs than many of us suppose." Throughout the book, Miklósi uses and discusses the use of terminology which has been developed to define the facets of the human–dog relationship, for example, paedomorphism, lupomorphism, doggerel, and my particular favorite "incanine" (inhumane), which I found most enlightening and helpful.

The first chapters explore early scientific exploration into canine modeling and difficulties encountered in terms of standardizing methodological approaches. If we want to understand more about behavioral malformations for example we must consider the functional importance of why dogs behave a certain way in certain situations. Tinbergen's four questions—function, mechanism, development, and evolution—are discussed, as they applied to early research. Miklósi calls for increased standardization of methodological approach which will enable the field to expand from a common platform of scientific thinking but incorporating researchers from diverse fields of archaeozoology, anthrozoology, genetics, ethology, psychology, and zoology.

There are discussions throughout the book about the similarities and, almost more importantly, the differences, between wolves and dogs. "Comparison of dogs to present day wolves, their closest genetic relative might be too restrictive because since the species split, modern wolves may have adapted to a different mosaic pattern of behavioral traits," whereas the natural environment of the dog is predominantly influenced by the homes and lifestyles of their owners. "Wolves in a home do not behave like dogs and dogs in a feral environment do not behave like wolves."

The schematic model of the dog population, based on data from the USA (p. 52), provides a simple but effective tool for between-country comparisons and highlighting the huge number of "unwanted" dogs passing through the surrender/ rescue shelter/ euthanasia process. Several studies which have demonstrated the benefits to humans of having dogs as companions, as assistance or working partners, are explored, but the benefits from the dogs' side of the relationship are strikingly absent. "Although dogs can physically hurt humans by biting, we can also hurt them if they are left to suffer in shelters." I would concur wholeheartedly with Miklósi's suggestions that "the darker side of human–dog relationships needs further attention" if we are to progress in the understanding and prevention of undesirable canine behaviors and ensure the welfare of the dogs with whom we share our lives.

Chapters 4 and 5 look at "a comparative approach to canis" and "domestication." For those interested in the ancestral origins of present day canids, Miklósi provides a comprehensive account of the research findings to date. He takes us through the last 40 million years of canidae evolution, documenting geographic shifts, archaeological findings, and genetic studies which help support or refute our current understanding of the co-evolution of humans and dogs. There are some fascinating discussions on the differences in hunting techniques, reproductive cycling, and parental techniques of feral dogs, wolves, foxes, coyote, and dingoes. He also uses an interesting analogy proposing that "the ancestors to dogs chose a novel anthropogenic niche, like an island, which offered unexploited resources and enabled them to enjoy reduced intraspecific and interspecific competition."

Having explored the methodological difficulties with the comparison of wolves and dogs, Miklósi then explores the influence of humans' manipulation of the wild canine genome, to create the diversity of the various breed types we have today, and how this further complicates our understanding of canine ethology. "Extreme relaxation of selection, when modern veterinary medicine enhances survival of individuals carrying deleterious mutations can increase the ratio of deleterious mutations in the population, especially when such dogs are not excluded from the breeding population." Our manipulation of the canine genome to create dogs with particular behavioral traits or appearance, whether for a particular utilitarian role or as companions, has slowed the evolution of the species by closing breeding populations and reducing the enhancing potential of inputting novel genetic material.

Neither an evolutionary, genetic, or archaeozoological approach can provide a full picture of the domestication process on their own, and again Miklósi calls for "further investigations which involve collaborative and refined methods for collecting data."

Chapters 6 and 7 explore "the perceptual world of the dog" and "physical ecological recognition." "Environmental stimulation can affect the survival of neurons which either centrally or in the sensory organ, determine the functional aspects of perception and as such the developmental environment of the dog will significantly influence its later perceptual abilities." Data are presented from studies which investigate vision, olfaction, and hearing in dogs, making a helpful comparison with the capabilities of humans. Differences between humans and dogs reflect the different functional requirements of each species; our visual acuity being 3–4 times better than dogs, whilst their hearing, olfactory abilities, night vision, and motion sensitivity far exceed that of humans. Practical application of both sensory and cognitive skills in training and working dogs are explored: for example, path following, searching for narcotics, spatial problem solving, and memory.

The chapter on social cognition provides further evidence on the differences between "independent, autonomous, problem solving wolves," versus dogs, "where the 'attachment relationship' predisposes the dog towards engaging in joint activities with human members of their group." Lupomorph, babymorph, and ethocognitive models are compared, with Miklósi favoring the latter which investigates a convergent evolution. Dogs are described as having a more "plastic behavioural phenotype" because their range of reactions in different environments is much larger than in wolves. This can result in problems if the environment does not provide the expected stimulation, resulting in behavioral differences or malformations (e.g., problem behavior in dogs). This becomes crucial when we consider the issues associated with canine aggression. Miklósi discusses current classifications of aggression and explains that ethological reasoning would show preference for "functional categories which recognize the targets of a contest." In relation to "potential threats" from humans, we do not know whether "dogs decoding the human signal rely on generalized information based on their species specific signals or whether learning plays an important role." "Improving our knowledge is important because of the misunderstanding in social communication based on the inappropriate signalling given by the human (especially children)."

The development of behavior in dogs is catalogued using the traditional developmental period model. Miklósi does, however, caution against the application of "critical periods" to the canine population in general because there is such strong evidence for breed specific variation in development and as such, puppy tests should be adapted for the specific breed in question. There are further comparisons between dog and wolf and the modification of canine development caused by the influence of humans: neonatal sensory development and influence of parenting; exploratory behaviors, play and feeding patterns; and reproductive and physical maturation times. Interestingly, "with wolves stimuli from humans in the neonatal period is effective only if exclusive and exposure to conspecifics has the potential to override this effect. Dogs socialized in a similar way show preference for the human if they are given the choice of a human or another dog instead." This effect is also mirrored in the chapter on temperament and personality, where Miklósi discusses a number of experiments which show the calming effect of humans on dogs and how they show a preference for human company in stressful situations, over that of another dog.

In the afterword, Miklósi concludes by saying "that recent changes in human living, which include lessening of social contacts and leading of a very individualistic lifestyle affect not only human relationships but also our relationship with dogs" and as a result "dogs are prevented from living a natural life in human communities because they spend most of their time alone or at the end of a leash." In his closing statement, Miklósi compares the role of dog ethologists to that of teachers and child psychologists "to teach humans in modern societies to keep up family life, which has always been essential for providing the appropriate social environment for both our children and our best friends."

I found Miklósi's approach to this book both honest and representative of current thinking on the origins and importance of canine ethology. Having said that, in light of his previous discussions on the influence and limitations inflicted on dogs by humans in relation to their habitat, behavior and genetics, I was surprised by the suggestion that the human–dog relationship should be considered as "a friendship." He has defined friendship as "alliance formation and cooperation and mutual social support, not excluding asymmetry, dominant or parental, in certain contexts, but including the possibility of leading an independent life and being an equal collaborative partner." I think that this idea alone would be worthy of further discussion and investigation, and I hope that, as Miklósi has suggested, we will see a multidisciplinary collaborative approach to further research which will improve our understanding of, and coexistence with, dogs so that the "friendship" can be more equalized for our future together.

When Species Meet

Donna J. Haraway. Minneapolis: University of Minnesota Press, 2008. 360 pages. ISBN: 978-0-8166-5046-0 (paperback); 978-0-8166-5045-3 (hardback)

Reviewed by: Randy Malamud, Department of English, Georgia State University, Atlanta, GA, USA. E-mail: rmalamud@gsu.edu

In her quintessentially eclectic and exuberantly discursive voice, Donna Haraway's *When Species Meet* explores the connections between people and other animals. The most prominent nonhuman animal in her discussion is *Canis lupus familiaris,* the domesticated dog, because, as she writes, "the familiar is always where the uncanny lurks" (p. 45). This book teases out the uncanny aspects of our familiar furry friends, foregrounding how much there is to deconstruct in our constructed social relationships with other animals. Fittingly, a dog named Ms. Cayenne Pepper, Haraway's intimately familiar companion, features resonantly. "Companion," Haraway reminds us, comes from *cum panis*, with bread, which provokes an interesting account in the final chapter, "Parting Bites," of people's gustatory interactions with other animals: connections are:

> [p]ropelled by the tasty but risky obligation of curiosity among companion species…the myriads of living organisms owe their evolved diversity and complexity to acts of symbiogenesis, through which promiscuous genomes and living

consortia are the potent progeny of ingestion and subsequent indigestion among messmates at table, when everyone is on the menu. (p. 287)

Traipsing through an array of theoretical engagements with animals, Haraway establishes her own praxis while grappling with others. Derrida, in his late work on cats, wins her approval for his awareness that his companion animal was indeed *someone*, and he "came right to the edge of respect . . . but he was sidetracked by his textual canon of Western philosophy and literature and by his own linked worries about being naked in front of his cat"; and so while he asked an important question ("And Say the Animal Responded?" was the title of his essay), he ultimately "failed a simple obligation of companion species; he did not become curious about what the cat might actually be doing, feeling, thinking, or perhaps making available to him in looking back at him that morning" (p. 20). He missed a chance to enter the world of this cat.

Haraway aspires to "become with" animals, but she demurs from Deleuze and Guattari's famous trope of becoming-animal. She lambastes the writers' "scorn for all that is mundane and ordinary and the profound absence of curiosity about or respect for and with actual animals" (p. 27), and she dismisses them in a huff: "No earthly animal would look twice at these authors" (p. 28).

Marx is useful in Haraway's enterprise, except that "he was finally unable to escape from the humanist teleology" (p. 46) of his analysis of labor, so Haraway extends his theory to a an imaginary project called "Biocapital," which extends the Marxian constructs of "use value" and "exchange value" by postulating an important third term, "encounter value." If we accept that "to be a situated human being is to be shaped by and with animal familiars," she writes, we "might deepen our abilities to understand value-added encounters" (p. 47), and this is what *When Species Meet* most importantly accomplishes: exploring the value that different species glean from interspecies encounters.

It's a complicated book, and a stimulating one. The rambling chapters are grounded in Haraway's personal and professional experiences, and soundly, if subjectively, extrapolated into challenges of conventional unexamined presuppositions about our relations with other species. Her discourse is compassionate, incisive, accusatory, nuanced, and sincere. There's a profusely detailed presence of animals—Australian shepherds, Iberian Churro, feral cats, chickens, among many others—and an elaborate account of the ethical, humanistic, and scientific internal dialogues that Haraway is having with herself (and with others, too, human and nonhuman, but the overall sense of her voice here strikes me as internalized ratiocination). She honestly records her initial conceptions and prejudices, and compellingly recounts her maturation of thought as her deliberations problematize her first instincts. For example:

In the beginning of everything that led to this book, I was pure of heart, at least in relation to dog breeds. I knew they were an affectation, an abuse, an abomination, the embodiment of animalizing racist eugenics, everything that represents modern people's misuse of other sentient beings for their own instrumental ends. …I was a true believer in the Church of the Shelter Dog, that ideal victim and scapegoat and therefore the uniquely proper recipient of love, care, and population control. …I have become an apostate. I am promiscuously tied with both my old and new objects of affection, two kinds of kinds, mutts and purebreds. Two terrible things caused this unregenerate state: I got curious, and I fell in love. Even worse, I fell in love with kinds as well as with individuals. Parasitized by paraphilias and epistemophilas, I labor on. (p. 96)

Companion animals are at the heart of this book, both as commodities and co-consumers. They are "patients, workers, technologies, and family members" (p. 62). They generate an enormous industry: $46 billion was spent worldwide in 2002 for pet care products, involving such sophisticated market-differentiation as vegan pet food, animal chiropractic practitioners, doggy Prozac, and so forth. As workers, dogs may be guides for the blind, psychotherapeutic aides for traumatized people, and rescuers in extreme environments, all of which involve interspecies training as a stage toward these "value-added encounters."

Another type of encounter Haraway explores is experimental, and she urges us to reject "unidirectional" and "self-satisfied calculation" (p. 71) about this sort of engagement between humans and other animals. She puts forth a model of "shared suffering," which takes place "in the midst of webbed existences" (p. 72), a tapestry of animal herds, sick children, research labs, neighborhoods, industries, economies, ecologies; this formulation encourages us to accept the responsibility of resisting the inherent inequality of animal lab research. Haraway weaves a philosophically complex approach to the ethics of experimentation that I cannot do justice to in this review, and I am not sure she concretely resolves all the problems she raises about how we may most responsibly address the subject; but certainly her concerns start us thinking in ways that add a fertile dimension to the established discourse.

A loosely-yoked mélange of subjects sometimes strains the reader's ability to connect the dots. A chapter on the sport of "agility" describes how dogs and human handlers work together to train and compete in contests involving jumps, tunnels, and dog walks. Haraway brings into this discussion issues of breeding, behavioral science, power, and place (in terms of the course ground's "contact zones" identifying the boundaries of competition, richly theorized to connect with Mary Pratt's analysis of how speakers of different languages meet in colonial encounters). In a chapter called "Crittercam," Haraway considers what it means when people attach optical technologies to animals and purport to render a posthuman medium of representation, an intensely authentic sensory experience. Like many of the other interspecies entanglements she considers, she completes her inquiry into this one with more questions than answers:

> There is no general answer to the question of animals' agential engagement in meanings, any more than there is a general account of human meaning making. …it's not about who "has" hermeneutic agency, as if it were a nominal substance instead of a verbal infolding. Insofar as I (and my machines) use an animal, I am used by an animal (with its attached machine). (p. 262)

When Species Meet is more philosophical and theoretical than programmatic or activist. It's sometimes difficult to follow, and sometimes seems inconclusive, though I presume that Haraway is aware of and untroubled by these rhetorical stances. "Nothing about the multi-species relationships I am sketching is emotionally, operationally, intellectually, or ethically simple for the people or clearly good or bad for the other critters," she writes (p. 281), and her book revels in this complexity.

WHERE THE WILD THINGS ARE NOW
Domestication Reconsidered

Edited by Rebecca Cassidy and Molly Mullin

Today, as genetic manipulation continues to break new barriers in scientific and medical research, we appear to be entering an age of biological control. Are we also writing a new chapter in the history of domestication?

From pet food to pigeon fanciers, *Where the Wild Things Are Now* provides an urgently needed re-examination of the concept of domestication against the shifting background of relationships among humans, animals and plants.

APRIL 2007 • 336pp • 15 b/w illus
PB 978 1 84520 153 1   £19.99 • $34.95
HB 978 1 84520 152 4   £55.00 • $99.95
www.bergpublishers.com

INDEX

List of Articles

Volume 21, Number 1

Social Effects of a Dog's Presence on Children with Disabilities
Stephanie Walters Esteves and Trevor Stokes — 5

Therapeutic Value of Equine–Human Bonding in Recovery from Trauma
Jan Yorke, Cindy Adams and Nick Coady — 17

Justifying Attitudes toward Animal Use: A Qualitative Study of People's Views and Beliefs
Sarah Knight and Louise Barnett — 31

Romantic Partners and Four-Legged Friends: An Extension of Attachment Theory to Relationships with Pets
Lisa Beck and Elizabeth A. Madresh — 43

Survival of Bottlenose Dolphin (*Tursiops sp.*) Calves at a Wild Dolphin Provisioning Program, Tangalooma, Australia
David T. Neil and Bonnie J. Holmes — 57

Go, Dog, Go: Maze Training AIBO vs. a Live Dog, An Exploratory Study
Aaron A. Pepe, Linda Upham Ellis, Valerie K. Sims and Matthew G. Chin — 71

Volume 21, Number 2

Psychological Sequelae of Pet Loss Following Hurricane Katrina
Melissa Hunt, Hind Al-Awadi and Megan Johnson — 109

Prosocial and Antisocial Behaviors in Adolescents: An Investigation into Associations with Attachment and Empathy
Kelly L. Thompson and Eleonora Gullone — 123

An Examination of the Relations between Social Support, Anthropomorphism and Stress among Dog Owners
Nikolina M. Duvall Antonacopoulos and Timothy A. Pychyl — 139

Equestrian Coaches' Understanding and Application of Learning Theory in Horse Training
Amanda K. Warren-Smith and Paul D. McGreevy — 153

Veterinary Students' Views Regarding the Legal Status of Companion Animals
François Martin and Sylvia Glover — 163

Embodying Anthropomorphism: Contextualizing Commonality in the Material Landscape
David Lulka — 181

Volume 21, Number 3

The Sympathetic Imagination and the Human–Animal Bond: Fostering Empathy
through Reading Imaginative Literature
Barbara Hardy Beierl — 213

Slovakian Pupils' Knowledge of, and Attitudes toward, Birds
Pavol Prokop, Milan Kubiatko and Jana Fančovičová — 221

Semantic Profiles of Zoos and Their Animals
Robert Sommer — 237

Comparison of Children's Behavior toward Sony's Robotic Dog AIBO and a Real Dog:
A Pilot Study
Filomena Nina Ribi, Akimitsu Yokoyama and Dennis C. Turner — 245

Relating Low Perceived Control and Attitudes toward Animal Training:
An Exploratory Study
Matthew G. Chin, Valerie K. Sims, Heather C. Lum and Mary Richards — 257

An Evaluation of a Pet Ownership Education Program for School Children
Grahame J. Coleman, Margaret J. Hall and Margaret Hay — 271

Development of an Item Scale to Assess Attitudes towards Domestic Dogs
in the United Republic of Tanzania
Darryn L. Knobel, M. Karen Laurenson, Rudovick R. Kazwala and Sarah Cleaveland — 285

Volume 21, Number 4

Politics, Press and the Performing Animals Controversy in Early Twentieth-Century Britain
David A. H. Wilson — 317

Domestic Dogs as Facilitators in Social Interaction: An Evaluation of Helping and
Courtship Behaviors
Nicolas Guéguen and Serge Ciccotti — 339

Attitudes and Actions of Pet Caregivers in New Providence, The Bahamas,
in the Context of Those of Their American Counterparts
William J. Fielding — 351

The Relationship between Childhood Cruelty to Animals and Psychological Adjustment:
A Malaysian Study
David Mellor, James Yeow, Norul Hidayah bt Mamat and Noor Fizlee bt Mohd Hapidzal — 363

Moral and Fearful Affiliations with the Animal World: Children's Conceptions of Bats
*Peter H. Kahn, Jr., Carol D. Saunders, Rachel L. Severson, Olin E. Myers, Jr.
and Brian T. Gill* — 375

Comparison of Vegetarians and Non-Vegetarians on Pet Attitude and Empathy
Brooke Dixon Preylo and Hiroko Arikawa — 387

BOOK NOTICES

Bears: A Brief History	197
Dog Behaviour, Evolution, and Cognition	197
Canis Africanis: A Dog History of Southern Africa	297
International Handbook of Animal Abuse and Cruelty: Theory, Research, and Application	297
Knowing Animals	85

Looking at Animals in Human History	85
Taxidermic Signs: Reconstructing Aboriginality	397
Victorian Animal Dreams: Representations of Animals in Victorian Literature and Culture	86
What Animals Mean in the Fiction of Modernity	297
When Species Meet	197
Why Dissection? Animal Use in Education	298
Wings in the Desert: A Folk Ornithology of the Northern Pimans	397

BOOK REVIEWS

Bears: A Brief History	401
Cat	91
Dog Behaviour, Evolution and Cognition	402
Fox	303
Ludzie i Ich Zwierzeta	89
Shaggy Muses: The Dogs Who Inspired Virginia Woolf, Emily Dickinson, Elizabeth Barrett Browning, Edith Wharton, and Emily Bronte	93
Six Legs Better: A Cultural History of Myrmecology	96
The Animals Reader: The Essential Classic and Contemporary Writings	301
The Moral Menagerie: Philosophy and Animal Rights	205
Victorian Fiction and the Cult of the Horse	203
What are the Animals to Us? Approaches from Science, Religion, Folklore, Literature, and Art	307
When Species Meet	405

CONFERENCES

Animal Athletes: Welfare of Animals in Sport	299
Animal Behavior Society meeting,	399
Canine Science Forum	86, 199
Experiential Learning in Humane Education	398
Human–Animal Interaction: Impacting Multiple Species	400
International Animal Welfare Conference	298
International Society for Anthrozoology	87, 200, 400
International Society for Applied Ethology	399
International Symposium on Animal-Assisted Therapy	200, 299
International Workshop on the Assessment of Animal Welfare at Farm and Group Level (WAFL)	87, 200, 299
Limits of Personhood	198
Minding Animals	200, 299, 399
Reflecting on Our Relationships: Animals and Agriculture	199
Research Center for Human–Animal Interaction	400
The Minds of Animals	86, 199
UFAW Animal Welfare Science Conference	86, 198
World Archaeological Conference	198

NEWS

ISAZ/WALTHAM Research Award	397
Section on Animal–Human Interaction: Research & Practice (AHI)	397

Subject Index

Animal Abuse
    associations with empathy and attachment, 123
    Malaysian children abusing animals, 363

Animal-Assisted Therapy/Activities
    effects of horses on recovery from trauma, 17
    social effects of a dog on children with disabilities in school, 5

Animal Welfare
    attitudes to animal use, 31
    horses in training, 153
    legal status of animals, 163
    performing animals controversy, 317

Anthropomorphism
    material factors affecting, 181
    relationship with social support and human stress, 139

Attachment
    association with prosocial and antisocial behaviors, 123
    human–horse relationships, 17
    theory, applicability to relationships with pets, 43

Attitudes
    of Bahamians and Americans to pets and petcare, 351
    of children to bats, 375
    of vegetarians and non-vegetarians to pets, 387
    of veterinary students, legal status of animals, 163
    to animals in literature, 213
    to animal training, 257
    to animal use, 31
    to birds, 221
    to domestic dogs in Tanzania, 285
    to performing animals, 317

Bats
    children's attitudes/reactions to, 375

Behavior
    abusive towards animals, Malaysian children, 363
    of adults to robotic (AIBO) and real dogs, 71
    of children to robotic (AIBO) and real dogs, 245
    of children with disabilities in school, effects of a dog on, 5
    of equestrian coaches in training horses, 153
    of people towards a person with a dog, 339

Birds
    attitudes to, 221

Children
    abusive of animals, Malaysia, 363
    attitudes/reaction to bats, 375
    attitudes to birds, 221
    behavior toward real and robotic dogs, 245
    conflict with sharing space with harbor seals, 181
    effects of pet ownership education program on, 271
    with disabilities, social effects of dogs on, 5

Companion Animals
    abuse of by Malaysian children, 363
    applicability of attachment theory to relationships with, 43
    attitudes to by vegetarians and non-vegetarians, 387
    attitudes towards dogs in Tanzania, 285
    Bahamian versus American interactions with/attitudes to pets, 351
    dogs as facilitators of social interaction, 339
    effects of a pet ownership education program on children's interactions with, 271
    effects of horses on recovery from trauma, 17
    psychological sequelae after loss, 109
    robotic (AIBO) versus real dog, human interactions with, 71, 245
    social effects of a dog on children with disabilities in school, 5
    veterinary students' attitudes to legal status of, 163

Cruelty, *see Animal Abuse*

Cultural Studies
    Bahamian versus American pet owners, 351
    Malaysian children and abuse of animals, 363
    Slovakian's knowledge of, and attitudes toward, birds, 221
    Tanzanian attitudes to dogs, 285

Dogs
- anthropomorphic behavior with, 139
- as facilitators of social interaction, 339
- as social support, 139
- attitudes to in Tanzania, 285
- effects of a pet ownership education program on children's interactions with, 271
- robotic (AIBO) versus real, human interactions with, 71, 245
- social effects on children with disabilities in school, 5

Dolphins
- survival of calves at a provisioning program, 57

Education
- effects of a pet ownership education program on children's interactions with dogs, 271
- social effects of a dog on children with disabilities in school, 5
- veterinary students' views on legal status of animals, 163

Empathy
- association with prosocial and antisocial behaviors, 123
- differences between vegetarians and non-vegetarians, 387
- fostering through literature, 213

Grief, *see Pet Loss*

Health
- effects of horses on recovery from trauma, 17
- human stress and dogs, 139
- psychological, effects of pet loss on, 109
- relations with social support and anthropomorphic behavior, 139

History
- performing animals controversy in twentieth-century Britain, 317

Horses
- coaches' understanding of learning theory in training, 153
- effects on humans in recovery from trauma, 17

Literature
- human–animal bond, effects on empathy, 213

Personality
- connotations of zoo animal species, 237

Pet Loss
- effects of on adults, 109

Pets, *see Companion Animals*

Politics
- performing animals controversy, 317

Representations
- animals in literature, 213
- performing animals in popular press, 317

Seals
- conflict with humans, 181

Training
- equestrian coaches' understanding of learning theory in, 153
- relationship with low perceived control and attitudes toward, 257

Wildlife
- harbor seals, conflict with humans, 181
- survival of dolphin calves at a provisioning program, 57

Zoos
- effects on children of bats in, 375
- semantic connotations of animals in, 237

Author Index

Articles

Adams, C., 17
Al-Awadi, H., 109
Arikawa, H., 387

Barnett, L., 31
Beck, L., 43
Beierl, B. H., 213

Chin, M. G., 71, 257
Ciccotti, S., 339
Cleaveland, S., 285
Coady, N., 17
Coleman, G. J., 271

Dixon Preylo, B., 387
Duvall Antonacopoulos, N., 139

Fan ovi ová, J., 221
Fielding, W.J., 351
Fizlee bt Mohd Hapidzal, N., 363

Gill, B. T., 375
Glover, S., 163
Guéguen, N., 339
Gullone, E., 123

Hall, M. J., 271
Hay, M., 271
Hidayah bt Mamat, N., 363
Holmes, B. J., 57
Hunt, M., 109

Johnson, M., 109

Kahn, Jr., P. H., 375
Kazwala, R., 285
Knight, S., 31
Knobel, D. L., 285
Kubiatko, M., 221

Laurenson, M. K., 285
Lulka, D., 181
Lum, H. C., 257

Madresh, E. A., 43
Martin, F., 163
McGreevy, P. D., 153
Mellor, D., 363
Myers, Jr., O. E., 375

Neil, D. T., 57

Pepe, A. A., 71
Prokop, P., 221
Pychyl, T. A., 139

Ribi, F. N., 245
Richards, M., 257

Saunders, C. D., 375
Severson, R. L., 375
Sims, V. K., 71, 257
Sommer, R., 237
Stokes, T., 5

Thompson, K. L., 123
Turner, D. C., 245

Upham Ellis, L., 71

Walters Estevez, S., 5
Warren-Smith, A. K., 153
Wilson, D. A. H., 317

Yeow, J., 363
Yokoyama, A., 245
Yorke, J., 17

Book Reviews

Brown, E. C., 96
Burt, J., 307

Cassidy, R., 301
Clutton-Brock, J., 91, 401
Copeland, M. W., 93
Corridan, C. L., 402
Cosslett, T., 203

Gruen, L., 205

Landry, D., 303

Malamud, R., 405

Szarycz, G. S., 89

NOTES FOR CONTRIBUTORS

The following details should be used **as a guide only**. Full details can be found at the Berg Publishers web site: www.bergpublishers.com.

Content

Anthrozoös will accept new contributions that describe the characteristics and consequences of interactions/relationships between people and non-human animals. Papers are welcome from the arts and humanities, behavioral and biological sciences, social sciences and the health sciences.

Types of Articles

Commentaries

This section provides a forum for raising issues relating to the fields of interest of the Journal, including theory, methodology, ethics, statistical analysis and nomenclature. Authors may make general points or provide critiques of particular published papers. These articles should usually be no longer than 5000 words.

Review Articles and Research Reports

Reviews—These should address fundamental issues relating to the interactions between people and other animals, and provide new insights into the subject(s) they cover. Original interdisciplinary syntheses are especially welcome. Reviews should be no longer than 8000 words.

Research Reports—both quantitative and qualitative reports are encouraged. These should cover subjects falling within the scope of the Journal and can be up to 6000 words in length.

Note: Word counts do not include tables, figures and references.

Manuscripts

Please submit an original and two other hard copies by post only (see mailing address below; electronic submission is **not** encouraged), along with a cover letter indicating clearly in which section you would like your manuscript to be considered. The two copies must **not** contain authors' names and addresses. After a manuscript has been accepted for publication, the author(s) will be requested to supply an electronic copy of it.

All text (including abstract, notes and references) must be typed (using, preferably, *MS-Word*), double-spaced, aligned left and printed on one side of each page only. Use active voice whenever feasible, and write in the first person. Tables and Figures should be in Helvetica or Arial font.

Use **American spelling and grammar conventions** throughout, except in non-American quotations and references.

Manuscripts should have line numbers and page numbers throughout. Authors whose first language is not English should have their paper checked by a native English speaker before submitting it.

Manuscript Organization

The title page of the original manuscript should contain the title of the article, and authors' names, affiliations, and present addresses. At the bottom of the page, give the full postal address, phone and fax numbers, and e-mail address of the corresponding author. In the following pages, provide an abstract (250 to 300 words), 3 to 5 keywords (in alphabetical order below the abstract), and the text, including, as appropriate, an introduction, methods, results, discussion, acknowledgements, notes, references, appendices, tables, figure captions, and figures. Each table/ figure must appear on a separate page. The authors' names should appear on the title page of the original manuscript only. Acknowledgements should be left blank until after the paper has been accepted.

Abbreviations and Units

Standard dictionary abbreviations are generally acceptable. Other abbreviations should be explained when first mentioned. SI units are preferred.

Footnotes

Footnotes appear as "Notes" at the end of articles. Authors are advised to include footnote material in the text whenever possible. Notes are to be numbered consecutively throughout the paper and are to be typed, double-spaced at the end of the text (do not use any footnoting or end-noting programs that your text software may offer, as this text becomes irretrievably lost at the typesetting stage).

References

For references in the text, give full surnames for papers by one, two or three authors, but only the surname of the first author, followed by "et al." for four or more (note that "et al." is neither underlined nor italicized). Check that all references in the text are in the reference list and vice versa, and that their dates and spelling match. Check foreign language references particularly carefully for accuracy of diacritical marks such as accents and umlauts.

Cite references in the text as, for example, Swabe (1998) or, if in parentheses, as (Daly and Morton 2006) or (McGreevy, Righetti and Thomson 2005). **Do not** use a comma to separate the author's name from the date. Where more than one paper by the same author has appeared in one year, the reference should be distinguished by "a," "b," "c," etc. (e.g., 1971a). If referring to a specific page in a book, please provide the page number in the citation: for example, (Serpell 1999, p. 45). Where multiple citations are referred to, place in chronological order, from oldest paper to most recent, using a semicolon to separate each reference: for example, (Harrison 1998; Gibbs 1999; Bekoff 2006).

The list of references should be arranged alphabetically by authors' names and chronologically per author. References cited with "et al." in the text should include all authors' names in the reference list. Journal titles should be given in full. References to books or monographs should include editors, edition and volume number, publisher, city and state or country where published, and relevant page numbers. A paper in press may be referenced if it has been accepted for publication. References to personal communications and unpublished work should appear in the text only.

Sample references (note: do not indent):

Galvin, S. and Herzog, H. 1992. Ethical ideology, animal activism and attitudes towards the treatment of animals. *Ethics and Behavior* 2: 141–149.

Lennon, R. and Eisenberg, N. 1987. Gender and age differences in empathy and sympathy. In *Empathy and its Development*, 195–217, ed. N. Eisenberg and J. Strayer. New York: Cambridge University Press.

Philo, C. and Wilbert, C. eds. 2000. *Animal Spaces, Beastly Places: New Geographies of Human–Animal Relations*. London: Routledge.

Paul, E. S. 1992. Pets in childhood: individual variation in childhood pet ownership. Ph.D. thesis, University of Cambridge, UK.

Tables

Each table must be presented on a separate page and be identified by a short, descriptive title placed at the top. Any necessary further explanations (e.g., the results of statistical tests) may be added as footnotes at the base of the table. Make sure that each abbreviation used in a table is fully explained in a footnote. Marginal notations on manuscripts should indicate approximately where tables are to appear. Please use Helvetica or Arial font for all tables. Each table must be cited in the text.

Authors using *MS-Word* or other word-processing programs must use those programs' table editors to create tables. **Do not** create tables by typing single lines of text followed by a hard return, with spaces or tabs used to align columns.

Figures

All illustrative material (drawings, maps, diagrams, graphs and photographs) should be designated "Figures" and must be cited in the text. For the review process, it is acceptable to supply photocopies of figures. Once a paper is accepted, the author will be required to supply high-resolution files/prints of figures (electronic files are preferred). Figures must be submitted as separate image files and NOT embedded in the word document. Figures will be reproduced exactly as provided. However, as they will be reduced in size to fit the Journal's page format, figures must be of a size which allow a reduction of 50%.

Criteria for Evaluation

Anthrozoös is refereed and papers will be accepted only after appropriate blind review. The general criteria for acceptance are that the research meet standards for publication in a specialty journal appropriate to its field and that it provide new information, sound hypotheses, or insightful analyses relevant to the content area of *Anthrozoös*. This is a multidisciplinary journal, and authors should be aware that their own discipline's jargon may be unfamiliar to readers from other disciplines. Please keep jargon to a minimum and provide a complete methods section. If you are in doubt about this, please err on the side of providing fuller explanations. The Editor can always cut material but cannot add it.

Copyright

Papers are accepted on the understanding that they are subject to editorial revision and that they are contributed only to this Journal. Copyright in the article, including the right to reproduce the article in all forms and media, shall be assigned exclusively to the Journal. The transfer of copyright to *Anthrozoös* takes effect when the manuscript is accepted for publication.

Proofs

One set of proofs will be sent to the corresponding author as an e-mail attachment (PDF). Only typographic errors may be corrected at this stage.

On publication, authors will be sent a PDF e-print (with nonprinting watermark) of the final, published version of their article for personal use, and will be able to order a free copy of the issue in which their article appears through Berg Publishers. Contact Ken Bruce at kbruce@bergpublishers.com.

Mailing Address

All manuscripts should be sent to:
Dr. Anthony Podberscek, Editor *Anthrozoös*
University of Cambridge
Department of Veterinary Medicine
Madingley Road
Cambridge CB3 OES, UK

Manuscripts will be acknowledged and entered into the review process (described above).

Queries about any of the guidelines can be sent to the editor via e-mail: alp18@cam.ac.uk.